William Thomas Belfield

**On the Relations of Micro-Organisms to Disease.**

The Cartwright Lectures...

William Thomas Belfield

**On the Relations of Micro-Organisms to Disease.**
*The Cartwright Lectures...*

ISBN/EAN: 9783337059590

Printed in Europe, USA, Canada, Australia, Japan

Cover: Foto ©berggeist007 / pixelio.de

More available books at **www.hansebooks.com**

ON THE

# RELATIONS OF MICRO-ORGANISMS TO DISEASE.

## THE CARTWRIGHT LECTURES,

DELIVERED BEFORE THE ALUMNI ASSOCIATION OF THE COLLEGE OF PHYSICIANS AND SURGEONS, NEW YORK, FEBRUARY 19, 21, 24 and 27, 1883.

BY WILLIAM T. BELFIELD, M.D.,

Lecturer on Pathology, and on Genito-Urinary Diseases (Post-Graduate Course), Rush Medical College, Chicago.

---

*Reprinted from* THE MEDICAL RECORD, *February and March*, 1883.

CHICAGO:

W. T. KEENER, 96 WASHINGTON STREET.

1883.

ON THE RELATIONS OF

# MICRO-ORGANISMS TO DISEASE.

## Lecture I.

*Mr. President and Gentlemen:* In accepting your flattering invitation to deliver the Cartwright lectures, I have, in compliance with your request, selected my present subject—almost the only one indeed which I would venture to discuss in your presence—not simply because of its intrinsic importance and interest, but also because there exists in the medical public of our land a diversity of opinion concerning it, which is not, in my estimation, warranted by the facts. For since trustworthy original investigations in this direction, demanding continuous devotion of the observer to the subject and the renunciation of other pursuits; demanding special training and experience; requiring laboratory and other expensive facilities; since such investigations, possible therefore in general only through State or corporate assistance, have been, and under existing circumstances must be made almost exclusively in other lands than ours; since, further, important results attained within recent years and published in foreign tongues have been as yet but partially incorporated in our standard literature; since the tendency of the practising physician—to which category we all, with rare exceptions, of necessity belong—is ever toward the cultivation of the art rather than the science of medicine; since, finally, there is a prevalent disposition to ignore the entire subject as trivial or fanciful; for these, and perhaps other reasons, there prevail, as it appears to me, some misconceptions as to the present state of knowledge on this subject. It shall be, accordingly, my effort to present in these lectures no original investigations, no theories nor views, but simply the facts already established, and the deductions incident thereto—an effort which I am encouraged to undertake by some familiarity with pertinent literature, and by some little practical knowledge of the methods and manipulations involved.

In order to discuss intelligibly the more recent and familiar subjects, such as the rôle of the bacillus tuberculosis, we must bear in mind certain facts, less sensational and perhaps less widely known, concerning the life-history of microscopic parasites.

Although even the early microscopists, beginning with Leeuwenhoeck (1675), observed and studied bacteria; although these minute bodies were observed in animals dead of septic infection by Fuchs, in 1848, and in the blood of sheep dead of anthrax by Brauell and Davaine, in 1849 and 1850, no effort appears to have been made to establish a genetic relation between the plants and the disease until the publication of Pasteur's work on fermentations, in 1861. Then the bacteria which had been the unenvied monopoly of biologists suddenly acquired deep interest for pathologists. The experimental work on septic infection, by Mayerhofer, Coze and Feltz, Rindfleisch, Waldeyer, and Recklinghausen, in 1865, 1866, and 1867, drew the attention of the medical public to the subject Meanwhile Lister, impressed with the results of Pasteur's work, and desperate (as I was informed by a Glasgow neighbor of his) at the death from pyæmia of several cases in rapid succession, anticipating the tedious progress of experimental science, submitted the question to empiric arbitration on the operating table. His clinical results revolutionized surgical methods on the one hand and infused new vigor into experimental pathologists on the other; the number of workers and of works so rapidly increased that to-day simple mention of the literature of this subject would be the work of hours. I deem it, therefore, inexpedient to attempt, in the limited time at my disposal, a historical sketch of the development of the question, and shall endeavor to present merely the present knowledge of the subject, with a review of the evidence upon which it rests.

Fungi—plants which, because devoid of chlorophyll and therefore unable to fabricate the necessary hydrocarbons, are usually parasitic upon animal or vegetable tissues, living or dead—may be divided into three general classes: 1, the mould fungi (aspergillus varieties, for example); 2, the budding fungi (*e.g.*, the yeast plant); and, 3, bacteria, or dividing fungi.

Bacteria were defined in 1875 by Cohn as "cells devoid of chlorophyll, of spherical, oblong, or cylindrical, sometimes twisted or sinuous form, which reproduce themselves exclusively by transverse division and live either isolated or in families." This definition requires to-day but one modification; reproduction, namely, is known to occur in several species, at least by the formation and liberation of spores as well as by fission. The bacteria consist of a nitrogenous, highly refractive, usually colorless substance, protoplasm or mycoprotein (Nencki), imbedded in which glistening, oily looking granules can sometimes be observed. The bacterium exhibits a power of resistance toward acids and alkalies which could not be possessed by a nitrogenous substance; hence it was assumed by Cohn that the external layer is a sheath of cellulose or similar hydrocarbonaceous material, such as the mould-fungi are known to possess. Although there has been as yet no decisive ocular demonstration of the existence of this membrane, the reactions of certain species toward straining fluids corroborate the assumption.

Many of the elongated bacteria have been demonstrated to possess also a thread like projection from the extremity, a flagellum or cilium; these, as well as some other varieties not yet proven to possess flagella, are capable of independent, often rapid locomotion in liquids; others are devoid of flagella, and incapable of motion; hence it is highly probable that the power of locomotion is associated with the possession of cilia. Beyond this bacteria seem to possess no differentiation of structure nor localization of function; nutrition and assimilation are processes of osmosis. This simplicity of structure and function has given rise to discussion as to whether they should be regarded as animals or vegetables; the question is, of course, merely a technical one of classification; since the features which distinguish the higher animals from the higher plants disappear as we descend the scale of organic life until few or none remain; yet because almost all of the simplest organisms hitherto called animals, the flagellata, possess a rudimentary mouth and are capable of absorbing solid food, while the simplest plants are not so characterized, the bacteria have been assigned to the vegetable kingdom. Whether they should be called algæ or fungi is a question for botanists to decide; the power of independent motion exhibited by some varieties suggests affinity with the algæ; but the absence of chlorophyll is generally considered to require their classification among the fungi.

The necessities of their existence are as simple as those of the mould fungi; indeed, so nearly identical as to require no discussion. As to the chemical reactions incident to their vital activity, our present knowledge is very scanty; one variety is known to induce the transformation of grape and milk sugar into lactic acid; another the decomposition of glucose or lactic acid with formation of butyric acid; another the change of urea into carbonate of ammonia; some produce pigments, blue, red, yellow; of many we know only that they transform a solid substance—gelatine, for example—into a liquid; but one of the most important facts in regard to them is the proof that putrefaction of albuminous substances is a phenomenon incident to the vital activity of certain varieties—the bacterium termo and probably others—as must be admitted by every one familiar with the work of Pasteur, Tyndall, and their pupils. Until the chemistry of their vital processes is ascertained, it will be impossible to assert how they can be injurious to a living tissue, whether by simple mechanical irritation, by the appropriation of oxygen and other nutritious elements, by the excretion of substances injurious to animal cells, or in several of these ways combined. The formation of substances incompatible with the life of the animal cells seems to play a prominent rôle in the production of injurious effects by at least some varieties.

The simplicity of organization and vital requirements explains their extensive distribution in nature: every moist substance of organic origin and all water containing even a trace of organic matter is favorable soil for one or more varieties; the upper layers of the earth, containing these essential ingredients, and remaining comparatively warm, constitute a continual breeding-place for these organisms. The minuteness and lightness of bacteria explain their presence in the atmosphere; they are swept by currents of air from dry or moist surfaces; they float in clouds of dust; they are carried by insects; the persistence of their vitality, the rapidity of their propagation, result in practical ubiquity. Direct microscopic observation of atmospheric dust, and the experiments of Tyndall with the electric beam in a dark chamber, have shown that wherever we find dust, at moderate temperature and altitude, we may expect to find bacteria. Yet the atmospheric bacteria are probably not so numerous as has been pictured. The observations of Miquel and of Koch show that even in a laboratory many litres of air contain no organisms. Whether or not bacteria are swept from surfaces of liquids; whether after once drying upon a given surface they can be removed by air-currents, are as yet undecided questions which may have practical bearings in the future. At present we know no laws of atmospheric distribution wherein bacteria exhibit other behavior than particles of dust in general.

The champions of spontaneous generation, compelled to surrender their maggots in decaying meat to the simple demonstration that covering the meat with fine gauze, which prevented the access of flies, prevented also the development of maggots; forced to abandon intestinal worms by the successive demonstrations of numerous observers that each worm proceeded, though often by devious ways, from a similar pre-existing organism; found a tower of strength in the bacteria, a position fortified by a series of careful, conscientious, and delicate experiments by Bastian and Pouchet. The result of that contest is known to all. The errors in manipulation and interpretation upon which the proof of spontaneous generation rested were detected, and a series of hitherto unassailable experiments by Pasteur, Tyndall, Traube, and Brefeld compelled the admission that bacteria, like the intestinal worms and the maggots, and all other living things, illustrate the dogma, "*omne vivum ex ovo!*" Yet this phantasy of spontaneous generation seems a spook which can never be exorcised from man's imagination. Quite recently Arndt has deduced from experiments, to which I shall presently refer, a conclusion which may be regarded as modified spontaneous generation; namely, that certain elements of animal-cells can, under favoring conditions, continue to exist and develop into bacteria after the death of the cells of which they were previously constituent molecules. Yet the evidence adduced does not as yet warrant any hesitation in accepting the current doctrine that bacteria, like all other organisms, proceed from pre-existent similar beings.

In order to discuss intelligibly the individual bacteria, we must agree upon a classification. The nomenclature has given rise to much discussion and more confusion. At first each investigator christened, after his own fancy, every new variety. The French school, first in the field by virtue of Pasteur's work on fermentation, employed, very loosely, the terms vibrios, monads, torulaceæ, etc. Natural selection has proven Cohn's classification the fittest to survive, which is quite natural, since the greater part of our exact knowledge of this subject is due to this distinguished botanist and his pupils. Cohn's original classification permits and will doubtless need amendments; in fact, he has already proposed some essential modifications based chiefly upon the form and mode of association of the individual cells. I shall adhere to that nomenclature in general use by the German mycologists.

Bacteria are distinguished in this system according to form simply into (1) micrococci, or spherical bacteria; (2) oblong bacteria, or simply bacteria; (3) bacilli, or rod bacteria; (4) spirilla, or spiral bacteria. A disadvantage in this nomenclature is the employment of the word bacteria to designate two different conceptions—the entire tribe including all four classes, a general name, and the second class oblong bacteria (Figs. 1, 5), in distinction from the others—a double signification which has led to

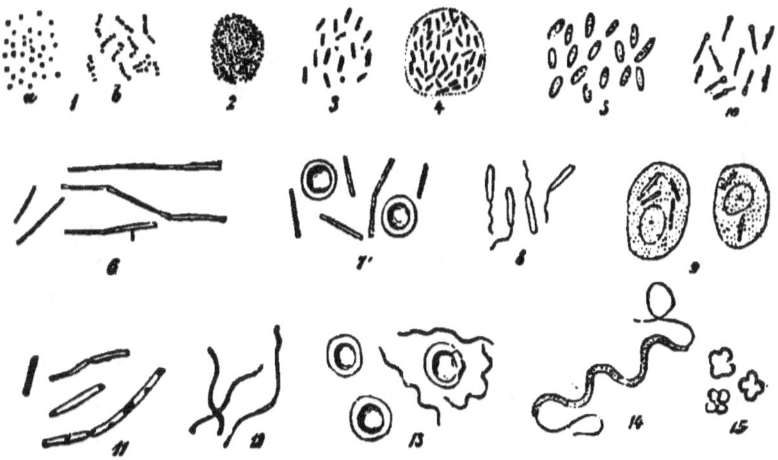

Fig. 1.—Various bacterial forms. 1. Micrococcus septicus; *a*, scattered; *b*, in chains—torula. 2. Same in zoöglœa form. 3. Bacterium termo. 4. Same—zoöglœa. 5. Bact. lineola. 6. Bacillus subtilis. 7. Bacillus anthracis and blood-corpuscles. 8. Bacillus (from mouth) with cilium. 9. Bacillus lepræ. 10. Bacilli with spores. 11. Bacillus malariæ. 12. Vibrio serpens. 13. Spirochæte Obermeieri. 14. Spirillum volutans. 15. Sarcina. × 500. (Copied from Ziegler's path. Anatomie, Jena, 1882.)

some confusion. The micrococci are the smallest, and, hence, individually least perfectly known; so small, indeed, often less than one micromillimetre in diameter, that nothing definite as to structure and contents has, as yet, been ascertained. They seem to exhibit in general no independent motion; they occur either isolated, in pairs, in chains (streptococcus or torula), or, when multiplying rapidly, in large numbers imbedded in a gelatinous material produced by the organism—the whole mass being termed zoöglœa. Some are colorless, others pigmented.

Of the bacilli, Cohn makes two genera—bacillus and vibrio; others add more varieties. The members of the genus bacillus are cylindrical rods; they multiply by fission, and some certainly exhibit a second mode of reproduction—the formation within their sheath of minute globular or ovoid bodies, spores, which subsequently escape by rupture of the sheath (Fig. 4), and are capable, under proper conditions, of growing again into the rod form. These spores exhibit a tenacity of life not possessed by the mature bacilli, nor indeed by other varieties of bacteria, since their vital activity is sometimes unimpaired by prolonged boiling, or by immersion for months in absolute alcohol, either of which procedures destroys mature forms. The spores seem under ordinary conditions the impersonations of immortality; time seems powerless to weaken them.

In order to study the phenomena accompanying the presence of bacteria in animal tissues, one must naturally first identify the bacteria. Now this is a more serious undertaking than the current literature on the subject in our language would imply. A bacterium is a mass of matter which possesses a definite size and shape, may or may not exhibit motion, has a certain chemical composition, and is capable of growth and reproduction—is, in short, a living organism; and no mass of matter can be justly called a bacterium until proven to possess these several characteristics; for one or more of these several properties may be exhibited by bodies found in the animal tissues which are not bacteria.

An unfortunately large number of publications on this subject exhibit by negative inference or positive demonstration, a failure to appreciate this self-evident fact. Size, shape, and, above all, movement, are considered conclusive evidence of bacterial nature. Size, shape, and presence or absence of motion may be determined by direct observation under the microscope; distinctive chemical traits may be detected by behavior toward certain reagents; the various aniline colors distinguish at once the protoplasm of bacteria from cell-bodies, fibrin threads, fat-granules, crystals; for every known bacterium absorbs one or more of the aniline dyes in watery or alcoholic solution. Yet size, shape, motion, and absorption of aniline dyes do not conclusively prove the bacterial nature of the body under examination, since the same phenomena may be exhibited by material forms which are not bacteria. Micrococci cannot thus be individually distinguished from the granules found in the nuclei of many cells, in leucocytes, and floating free in the blood; rod bacteria are sometimes closely simulated in size and shape by fibrin threads and organic crystals. Groups of bacteria, especially of the micrococcus tribe, are simulated by nuclear detritus floating in the blood, as pointed out by Riess; by large granular cells, the "Mastzellen" of Ehrlich, which are found in large numbers in various inflamed tissues, in diphtheria, typhoid fever, elephantiasis Græcorum, for example; by cross-sections of fibrin threads in blood-vessels; by globular masses usually considered to be leucine, which may occur apparently in any tissue, in normal as well as in certain pathological states. It is true that experience teaches one to distinguish these bodies from bacteria, in some cases, by their appearance and reaction to staining agents. Yet absolute certainty can usually be secured, even by the experienced mycologist, only by cultivation outside of the body. In the independent exhibition of reproductive power, by fission or spore formation or both, lies, therefore, the only positive proof that a particle under examination, exhibiting the size, shape, and reaction to staining agents characteristic of a bacterium, is actually one of these lowly organisms, and not an unorganized mass of similar appearance.

The theoretical considerations end here, but the practical difficulties begin; for, in order to be sure that an organism which grows in a liquid outside of the body is the same as the particle previously observed within the tissue, we must be assured that no other organism can have obtained access to the culture fluid; for the microscopic dimensions of the particle prevent continuous observation during the transfer, and the morphological similarity of different varieties, especially among the micrococci, render individual recognition impossible. Every object which can come into contact with the liquid or the particle under examination—the skin of the animal from which the tissue is transferred, the instruments, the vessel or slide containing the nutrient material, the material itself, the surrounding air, so far as possible—must be sterilized, liberated from all contained and adherent organisms; and even then there remains an element of uncertainty, since it is impossible (by the ordinary methods of cultivation) to demonstrate that these precautions have been efficient. These difficulties of sterilization cannot be fully appreciated without actual experience, which soon demonstrates that the greatest care and attention is often impotent to secure the isolation of a given species from other bacteria; Bastian's famous experiments in support of spontaneous generation may serve as an illustration.

In his earlier work, though carefully and conscientiously performed, the apparently spontaneons appearance of bacteria in various animal and vegetable infusions was easily explained by his failure to previously heat the glass vessels in which the infusions were kept; for the observance of this now elementary precaution prevented the appearance of organisms in the liquids. But later experiments seemed indeed unassailable; he found that thoroughly boiled urine remained, in a previously heated and well-stoppered flask, perfectly free from organisms; when however the urine was made alkaline by the addition of a caustic potash solution, also previously boiled, the conditions remaining otherwise unchanged, bacteria were soon developed in immense numbers. Bastian explained by the hypothesis that alkaline was more favorable than acid urine to the generation of these organisms. Pasteur, skeptical as to the accuracy of Bastian's manipulations, repeated the experiment and secured the same result—bacteria were developed, even with the greatest possible attention to details of execution. He found, however, that if the caustic potash were added to the urine, not in watery solution but in the pure state after heating to redness, no organisms were developed; further, that if Bastian's solution of caustic potash and urine were heated to 110° C., no development of life occurred. The error was therefore the assumption that all organisms in the potash solution, as well as in the urine, were destroyed by boiling—an assumption now known to be at variance with the fact. Tyndall also narrates an instance which might have been in less careful hands misinterpreted. He removed from a clear, sterilized infusion a drop of liquid, and to his astonishment found it, under the microscope, swarming with bacteria; examination of a second drop showed none. The mystery was soon explained; he had cleansed his pipette before taking the first drop with distilled water, a drop of which had remained in the tube, and which contained, as examination of the water in the bottle revealed, numerous bacteria. In these days pipettes are cleansed, not with distilled water, but by a Bunsen flame; knives, needles, test-tubes, flasks, etc., are considered sterilized after heating for five minutes to 150° C.; and fingers are allowed under no circumstances to touch anything which could possibly come into subsequent contact with the culture.

A consideration of the difficulties thus briefly sketched in the way of even the accurate recognition of bacteria, discloses the value to be attached to many publications concerning these organisms—which exhibit but too often the author's neglect to comply with, sometimes even his ignorance of, the elementary requirements of principles and practice. As illustrations I shall select from the mass of literature of this description a few which, from the eminence of their respective writers in other departments of medicine, have attracted considerable attention without as well as within our professional ranks. Some four years ago a most genial and accomplished gentleman, an eminent practitioner of New England, made a tour of our large cities with the benevolent object of instructing his professional brethren as to the etiology of tuberculosis and of syphilis. He asserted the discovery in the blood of such patients of "germs" not found in other blood; he exhibited a series of transparencies in which granules—the germs—were shown in the blood; he showed a loaf of bread which had been fermented by organisms from the dejections of tuberculous and syphilitic patients. The good work was not limited to the medical profession; a noted Boston clergyman delivered a Monday lecture upon this subject, in which the doctor's blood-granules, projected upon a screen, posed before the conscience-smitten audience as the avengers through whose dread agency the way of the amorous transgressor sometimes becomes hard. Only two links were lacking in the chain of evidence; first, there was not the slightest proof, nor attempt at proof, that the blood-granules were germs; second, it was unfortunately demonstrated that identical granules were usually found in healthy individuals.

A few weeks ago the startling discovery was announced that the famous bacillus tuberculosis was a fat-crystal. A distinguished pathologist, failing to detect the bacilli in tuberculous tissues, treated his sections with a thirty per cent. solution of caustic potash, whereupon crystals of fat of course appeared. These crystals are alleged to be identical with the bacilli tuberculosis; therefore these organisms exist only in the imagination of benighted individuals who blindly follow Koch.

Now what proof is adduced that the crystals and the alleged bacilli are identical? Merely that they have the same size and shape. When we remember that the discoverer of the crystals had never seen the bacteria in question, we can admit that the former have the same size and shape as the bacilli of Dr. Schmidt's imagination, but not necessarily as the actual organisms. A comparison of the two demonstrates, as Dr. Schmidt has recently learned by personal observation, that they are widely different, even in appearance. But, assuming that in size and contour Schmidt's crystal and Koch's bacillus were similar, would that justify the assertion of their identity? Evidently not to an individual whose conception of a bacterium comprises something more than size and shape. The crystal cannot be made to absorb and retain aniline dyes, as Dr. Schmidt expressly states; the bacillus, like other bacteria, is readily stained by any one of several aniline colors. The crystal, we may assume, does not grow nor reproduce; the bacillus elongates, divides, and produces in its substance two or more globular bodies, which in turn grow into rods. Dr. Schmidt fails to appreciate these vital differences; ignores the absorbent and reproductive powers of the true bacilli, attested by a score of competent observers. For him size and shape are enough, and upon this fancied resemblance of his crystals, in outline, to bodies that he has never seen, he assumes their identity. I shall consider Schmidt's paper in the discussion of tuberculosis, and make this allusion here because it illustrates admirably the fact that even in these latter days publications about bacteria, by men of experience in other departments of medicine, even in pathology, evince a failure to appreciate the first principles of mycological investigation.

To ascertain the relation to a disease of bodies whose bacterial nature is thus recognized, and which are found in the blood or tissues of a diseased animal, it is evident that the first step must consist in the perfect isolation of the bacteria from the enclosing tissue; since otherwise the possible effect of inoculation may not be ascribed to the bacteria rather than to the accompanying unorganized substances.

The isolation of bacteria from the blood and tissues of an animal has been attempted chiefly in two ways: by simple filtration through paper, clay, or other porous substance, and by artificial cultivation. Filtration is evidently a very unsatisfactory attempt at isolation; the separation of the smaller bacteria, especially micrococci, is practically impossible; and since other ingredients of blood or tissue are or may be retained on and in the filter, the proof of isolation is not convincing. In artificial cultivation advantage is taken of the fact that the bacteria reproduce indefinitely, while the animal tissues of course do not. A drop of blood or pus, or a particle of tissue containing the bacteria, is placed in an appropriate nutrient medium in a flask, tube, or other receptacle. In a few hours or days the organisms have become diffused, by virtue of their rapid multiplication, throughout the entire liquid. A drop, or fraction of a drop, of this fluid containing bacteria, is then transferred to a second vessel similarly prepared; after the growth of the organisms in this, a minute quantity is transferred to a third flask, and so on indefinitely at the will of the operator. In this way the bacteria can be practically isolated from the animal tissues introduced with them into the first culture vessel; and the effect of inoculation from the tenth, twentieth, or thirtieth successive culture cannot be reasonably ascribed to the unorganized constituents therein contained. But it is evident that in order to attribute the effect of such inoculation to that particular bacterial species contained in the diseased animal, one must be absolutely certain that no other variety has obtained a foothold in the cultures—that the original bacteria are isolated not only from the animal tissues—a comparatively simple matter—but also from all the other varieties of bacteria—which seem omnipresent. And just here is the difficulty which has been until recently almost insuperable; here is the possible source of error which weakens materially some brilliant deductions from experimental work; and this possibility of error is the basis of the general criticism which Koch urges against Pasteur's work—a criticism which, as is evident from a comparison of methods, is not without foundation. Although numerous modifications of culture methods have been employed, all may be grouped in three general classes. The first, the earliest, and by far the worst, is cultivation in flasks, tubes, or other vessels containing the nutrient liquid, usually in large quantities—one or more ounces. This method, thanks to its adoption and retention by Pasteur, as well as to the facility of its manipulation, has secured an unfortunately extensive and persistent employment. The defects become apparent when we consider the vital properties of bacteria. Different species require different pabulum and various temperatures for their successful cultivation. The bacillus subtilis (Cohn), for example, grows luxuriantly in a simple infusion of hay, which is ordinarily slightly acid in reaction; the bacillus anthracis, which is morphologically similar, indeed almost identical with the former, grows very slowly or not at all in the same infusion; the addition of a little magnesia or other base, sufficient to render the liquid somewhat alkaline, reverses the relative reproductive activity of the two. The hay bacillus (b. subtilis), again, can reproduce at a temperature incompatible with the reproduction of the b. anthracis. If, then, the two be sown in the same sterilized hay infusion, the crop will be determined largely by the reaction and by the temperature of the liquid. Another feature to be remembered is the variable rapidity with which different species multiply, even under the most favorable surroundings; for as Nägeli has shown, if two bacterial varieties, A and B, be present in a liquid adapted to each, A dividing its cell into two in twenty-five minutes, B in forty minutes, the latter, even if present at first in 1,800,000 times the number of the former, will in eighty hours have been stifled by A. In a mixed cultivation—in other words—the quickest to propagate will, *cæteris paribus*, in a few hours or days remain alone—a principle with whose applicability to higher organisms we are of course familiar. It is evident, then, that the best method of isolation is that which affords (first) the greatest security against the intrusion of other organisms than that under cultivation, and since such intrusion cannot by any method at present employed be with certainty prevented (second), the greatest probability of the detection of such invasion by other bacteria. The isolation cultures in flasks suffer the same dangers of adulteration as other methods, aggravated somewhat by the increased difficulty of sterilizing large quantities of the nutrient material. These dangers recur with the institution of every new culture; for in order to secure isolation of the organisms from the blood or tissue which accompanied them into the first flask, many transplantations must be made. Successful induction of infection by inoculation from the first culture is of course, met by the objection that the liquids contains solid or fluid unorganized ingredients from the animal diffused through it, and that the effect cannot be attributed to the contained bacteria alone.

With every transfer to fresh culture fluid the possibilities of adulteration by intrusion of other organisms contained in the liquid, the flask, the air, etc., are encountered; moreover, the last culture, however long the series may be, is theoretically, at least, a dilution of the animal juices present in the first flask, although by many successive generations the dilution becomes homœopathic and may be practically disregarded. That this danger of the intrusion by other bacteria during transfer, however carefully done, is not trivial is known to every experienced observer. Mr. Cheyne ("Antiseptic Surgery," page 261) says: "In the room in which I work I have never been able, without the aid of the spray, to transfer micrococci from one flask to another. For in the latter flask bacteria" (in the generic sense) "almost invariably developed."

The chief objection to flask- or tube-cultivations, however, one which renders them utterly unsatisfactory as attempts at isolation, is the impossibility of detecting with certainty the presence of foreign organisms. Some varieties, it is true, indicate their presence macroscopically, but the absence of them does not prove the absence of others. One is compelled to remove the cotton, withdraw a drop of the liquid, and submit it to microscopic examination—a proceeding perilous to the purity of the culture. But even in this way no certainty can be assured, for the one drop may be free from intruding organisms, which may nevertheless be present in the flask. More than that; since many varieties, at least many bacteria growing under different circumstances, are morphologically indistinguishable, one is not always certain that the organisms found are really the offspring of those planted rather than morphologically identical intruders. Suppose, for example, we are attempting to isolate the micrococci found in the blood of a pyæmic patient; on examining a drop from our tenth generation, our tenth successive flask, we find only micrococci identical in appearance with those planted. Are we warranted in asserting that these are the descendants of the organisms seen in the blood? To answer the question affirmatively we must assume first that the micrococci contained in the blood were the only ones which gained access to the first flask; then that our ten flasks, ten liquids, ten stoppers were rendered and remained sterile; that our pipettes, forceps, etc., have been always free from organisms, and that during the nine transfers, and perhaps numerous test examinations, no micrococci have gained access to our cultures from the air—an assumption not warranted by experience. By such assumptions Bastian demonstrated spontaneous generation.

A source of possible error in these methods, not always recognized, is the necessary assumption that apparent community of form among bacteria proves identity of function. The fallacy of such assumption is *à priori* evident, and has been repeatedly demonstrated. A Cyclops whose exaggerated retina might fail to distinguish objects less than three feet in diameter, could perceive no morphological characteristics wherein a wolf differs from a sheep, and could have no conception of the perversity which distinguishes a mosquito or a wasp from the house-fly. In 1875 even Cohn pronounced the harmless bacillus subtilis of hay infusion morphologically identical with the bacillus anthracis; to-day we can, thanks to improvements in technique, distinguish the latter not only from the hay bacillus, but also from two other important varieties which are morphologically extremely similar, though physiologically quite distinct. Further research may possibly reveal differences of form even among the minute bacteria. But granting that two micrococcus varieties are actually, as well as apparently, identical in form, that alone proves identity of function as conclusively as the morphological identity of two sugar-coated pills, containing arsenic and bread respectively, establishes the unity of their effects.

I have considered this subject at such length because it will be necessary, in the course of these lectures, to criticise certain experiments, and reject as not proven certain conclusions, because attained by this method of cultivation in liquids. The solution of many problems would be, in my judgment, materially hastened, and much conflict and confusion avoided by the substitution for this still popular method of one demonstrably far more accurate, to be presently described. Flask- and tube-cultures are certainly convenient for cultivation *en masse*, but evidently imperfect for investigations as to the rôle of bacteria in disease.

The fact that bacteria growing under different conditions, and apparently functionally different, are morphologically identical and all so widely distributed and ubiquitous, renders it evident that one condition must be fulfilled in order to demonstrate conclusively and finally that the organisms present in a culture at a given moment are the progeny of those previously planted by the experimenter; the former must be observed *to proceed by continuity of structure*, from the original bacteria. In this way, and thus only, can all doubt and objection be silenced; all possibility of misinterpretation be eliminated. The first requisite for the execution of this method is evidently continuous observation of the growing bacteria through the microscope; a method long used. A drop of nutrient liquid is placed on a cover-glass, the bacteria sown therein, and the cover inverted over a hollow slide or a cell of glass cemented to the slide; the edge of the cover is then or previously smeared with oil, for the double purpose of limiting evaporation from the droplet and of preventing the intrusion of foreign substances or organisms.

If foreign organisms capable of growing in the liquid obtain, by any error of manipulation or other means, access to the droplet, they may become in a few hours diffused throughout its extent; if perceptibly different in size or shape from the variety sown, the intrusion may be detected and the culture abandoned; if morphologically indistinguishable, the intrusion cannot be detected, the culture is considered pure, and lamentable errors of interpretation may result. Even in Koch's skilful hands nearly one-half such cultures were found to contain foreign organisms.

Such, then, were the methods of attempted isolation in general use until three years ago, and unfortunately still extensively employed; methods which, even in skilful hands gave conflicting results, and in inexperienced hands have demonstrated incredible things. By these methods spontaneous generation is easily proven; the metamorphosis of a pathogenic into an innocent bacterium, and conversely, is established with great facility; in fact almost any plausible hypothesis can, with a reasonable amount of ingenuity and inexperience, be clearly demonstrated. It is difficult to conceive, indeed, a shorter road to scientific notoriety than by the cultivation of bacteria in flasks—an opportunity already amply improved. One need only extend his series of successive cultures long enough, make transfers and pipette examinations often enough, permit the temperature of his oven to vary a little, perhaps spice his work with a touch of the fashionable evolution hypothesis by varying the composition of his culture-fluids so as to invite the growth of various organisms, and, presto! there appears a conclusion whose adoption would revolutionize mycology, medicine, and perhaps modern society. A bacillus may be made to grow out of a bacterium, this out of a micrococcus, or any of them out of nothing. A bacillus anthracis can, in the course of fifteen hundred transfers, be robbed of its terrors—the leopard can change its spots, the lion can become a lamb.

Three years ago there was introduced—thanks to the

ingenuity of Koch—a method which avoids, theoretically and practically, the difficulties inseparable from previous attempts at isolation of a given bacterial species found in an animal, both from other varieties and from the accompanying animal juices. The essential feature of this method consists simply in the substitution

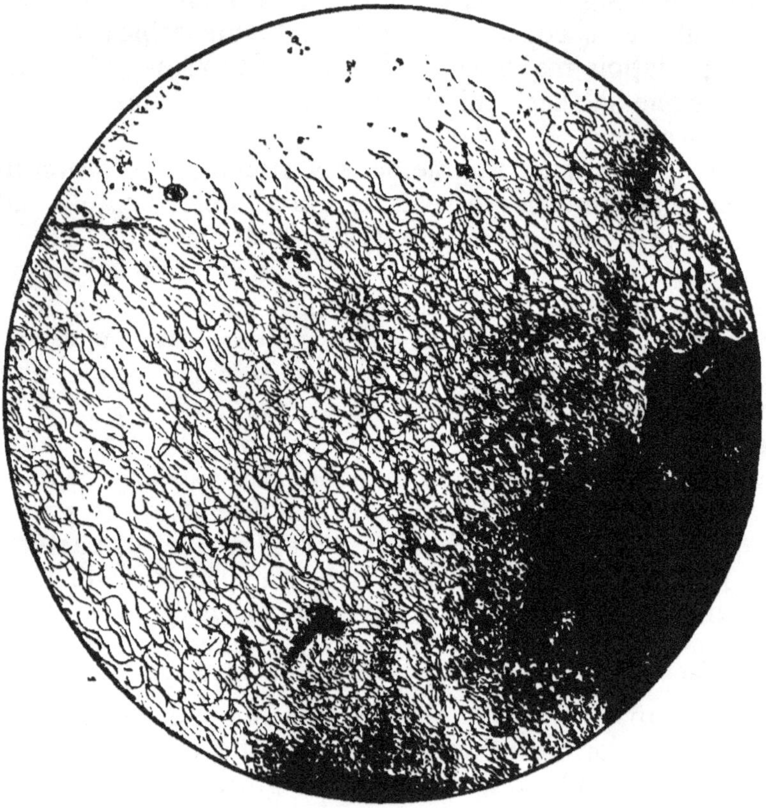

Fig. 2.—Anthrax bacilli growing from rabbit's liver, on blood serum. × 40. (This and succeeding cuts, illustrating Koch's method, are copied from original photomicrographs exhibited at the lecture.)

of a solid—transparent when possible—for the liquid material adapted to the nutrition of the organism.

The general plan is as follows: a solution of gelatine, beef-extract, pepton; or blood-serum, the relative proportions of the various ingredients varying with the species to be cultivated, is sterilized by repeated heating and then spread as a thin layer upon a disinfected slide and allowed to dry or coagulated by heat. A previously heated needle or scalpel is then dipped into the material containing the bacteria—septic mouse-blood, for example—and drawn lightly over the surface of the culture substance on the slide, or a series of punctures are made with the point.

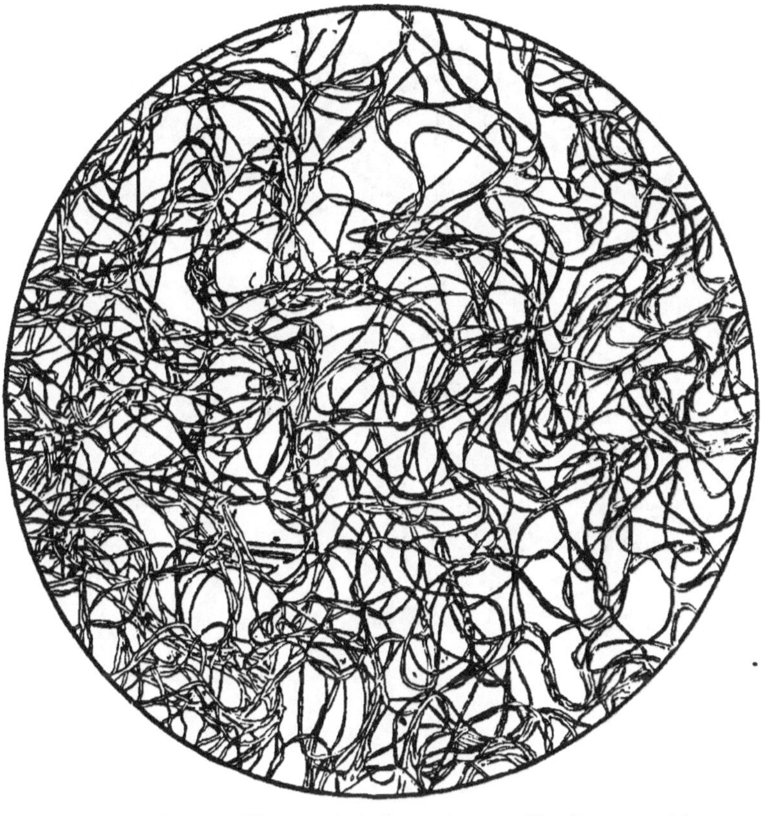

Fig. 3.—Anthrax bacilli, pure (isolation) culture on blood serum. × 140.

Thus a number of shallow furrows may be made, in and on the edges of which the bacteria are deposited. The slides so prepared are then transferred to the incubator or placed under a bell jar; or, if they are to be long preserved, in a thoroughly disinfected vessel closed with cotton. A disadvantage of this method, like that with liquid media, is the uncertainty of sterilizing the nourishing medium; further, the occasional settlement and growth of foreign organisms in the vicinity of those planted. Yet these accidents are in this method, in distinction from others, detected with ease and certainty (Fig. 5). For there is no diffusion through

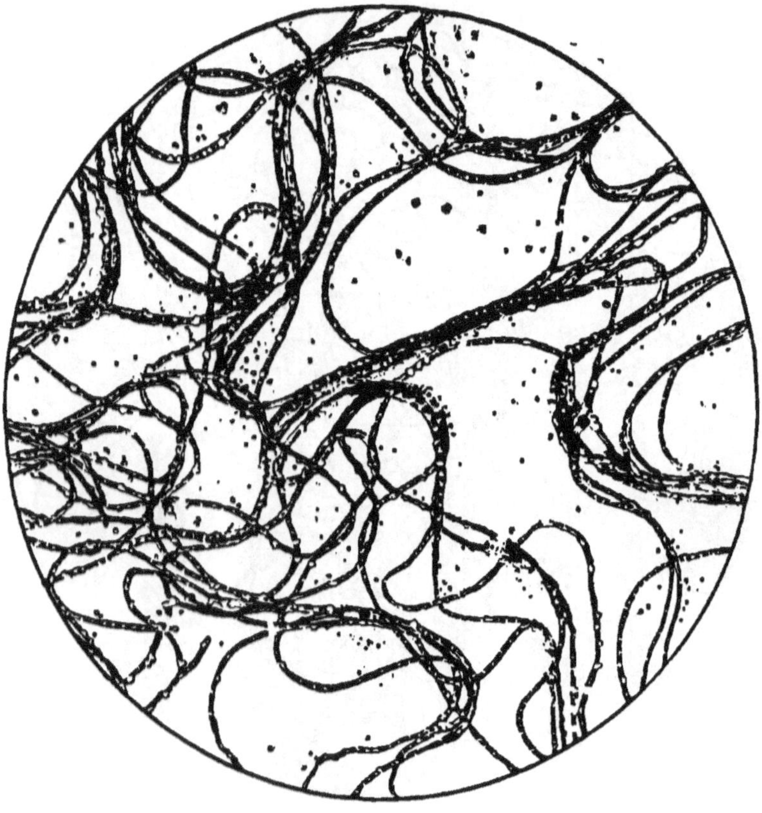

Fig. 4.—Anthrax bacilli—showing spore formation. × 300.

this solid medium of growing bacteria; each organism, whether intentionally planted or not, remains in one spot, any admixture must occur by continuity of growth.

By this method, then, there is ocular demonstration that the organisms of the successive cultures proceed as continuities of structure from those of the first; we may see the bacteria in the leucocytes of septic blood freshly deposited on the first slide, watch them multiply until they break out of the containing cell and extend over the gelatine, transfer these to a second slide, observe their continued multiplication, and so on indefinitely. By this method, and by no other at present employed, we have

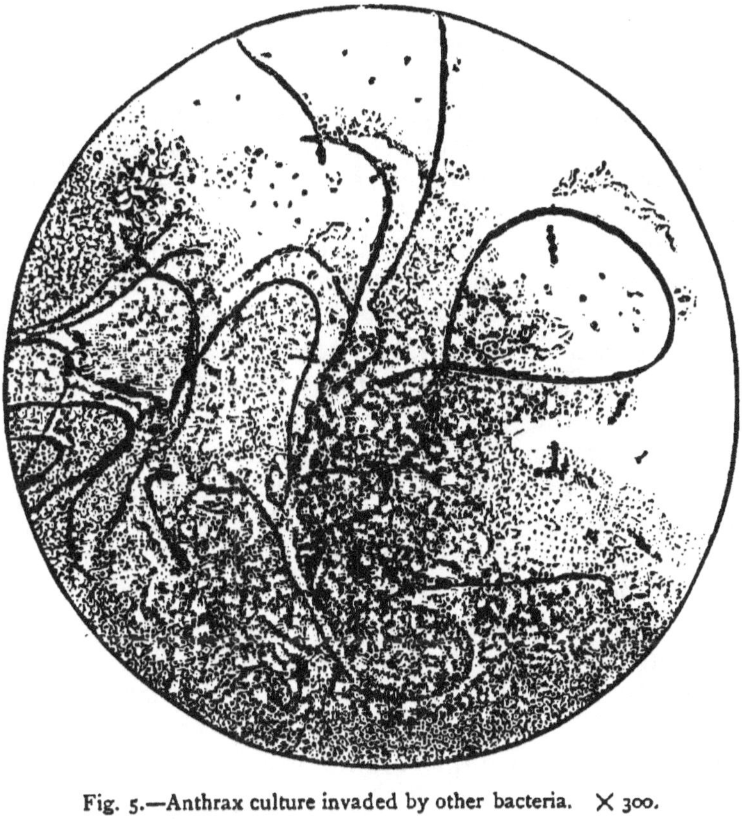

Fig. 5.—Anthrax culture invaded by other bacteria. × 300.

ocular demonstration that the organisms injected into an animal from the tenth, fiftieth, hundredth culture are structural continuities of the organisms present in the first animal; by this method we know that the tenth culture, for example, contains absolutely nothing of the original tissue except that incorporated into the living organisms; by this, and by no other method with which I am acquainted, it may be unhesitatingly affirmed that the effect induced in a healthy animal by inoculation with the tenth, twentieth, or fiftieth culture must be ascribed not only to bacteria, but to the direct descendants of those bacteria contained in the animal from which the first cultures were made. And permit me to add that after such demonstration has been furnished in scores of repetitions with identical results, by independent competent observers, by every one indeed, who has attempted it, it is unwise and unreasonable, to put it mildly, for any of us to deny, oppose, ignore, scoff at, or equivocate about a fact established far more firmly than are the majority of accepted facts, so called, in the science and art of medicine.

A consideration of importance in estimating the value of observations upon bacteria, is the microscopical technique employed by the observer. That good objectives should be used is well understood; but it is now essential also, in delicate investigations, to employ a condenser made after the pattern of Abbé's illuminator. The feature which makes this apparatus invaluable in the detection of bacteria is the possibility of obliterating, through the large aperture of the lenses, the picture of the object due to refraction, thus practically eliminating from the field of vision everything which is not colored. Bacteria not only absorb coloring matter—aniline dyes, for instance—but retain these pigments in the presence of reagents which decolorize the animal tissues, generally speaking. To detect a minute organism, the bacillus tuberculosis, for example, amid the cells, fibres, and granules of lung-tissue is, with ordinary illumination, extremely difficult or impossible; but after decolorizing the tissue and then, by means of the illuminator, obliterating refraction outlines, the bacilli appear almost alone, by virtue of their retained color, in a luminous field—distinct and easily recognizable. When, therefore, an observer asserts the absence of all bacteria from a tissue—tuberculous or leprous, for example—usually supposed to contain them, it is in order to inquire first whether he knows how to use the aniline dyes, and second, whether he has the optical means, an Abbé's or similar illuminator, necessary to enable him to utilize these staining methods in the detection of bacteria; if he lack either he is but ill-equipped for public entrance into the arena of original investigations on this subject; for both are essential to the development of the histologist into the medical mycologist.

I have thus endeavored, Mr. President, to sketch the principles involved in the study of bacteria, in order that we may on subsequent occasions form a somewhat accurate and intelligent estimate of the evidence already adduced as to the relations of these minute beings to various morbid conditions of the human subject. If the effort to convey the facts has been successful, you will agree that an original contribution to this evidence demands serious consideration only when it evinces, on the part of its author, not simply skill and experience as a microscopist and histologist, but also a practical acquaintance with and employment of the special methods and manipulations necessary for the recognition and cultivation of bacteria. Medical mycology has become a special department of investigation, comprising most delicate and easily vitiated technical methods, to be acquired by special study and experience; the time has passed when the ability to focus a quarter-inch objective entitled a man to an opinion upon the subject. Medical mycology may not be compared in age or attainments with chemistry; yet the necessary technique is as delicate in the one as in the other; and there exists therefore the same necessity for the recognition of individual knowledge, and by consequence of general ignorance of the one subject as of the other. I am led to indulge in these rather commonplace remarks by the observation that there is a more or less prevalent lack of appreciation of the evident facts in the case; witness the recent discovery of the pseudo-bacillus tuberculosis and the comments thereupon in current literature and society meetings.

Not a few medical journals announced the complete annihilation of the bacillus tuberculosis in particular and bacteria in general, a dictum pronounced by a witty rural editor, equally familiar with mycology and Latin syntax, in the words—"Sic transit bacteria." Even a leading New York journal remarked that "the case against the bacilli is very strong." If it should be announced by Dr. A. that pyæmia is not characterized by an excess of carbolic acid in the urine, as asserted by Brieger, we practising physicians would not undertake to decide; we know too much about chemistry. There are at present perhaps a score of men who have given abundant evidence of competence in bacterial investigations; to them and to such as they, not to dermatologists, surgeons, or pathologists, we must look for facts upon this subject, and for experimental criticisms of one another's assertions.

Before discussing the relations of bacteria to the body in disease, it is well to consider their relations to the animal in health and after death. It is a much disputed question whether any varieties of bacteria may exist in the blood or tissues of a healthy animal. In former years the affirmative was maintained by many, notably Billroth; with improvement of methods and differentiation between organized and unorganized particles the number of such affirmations decreased; and the three most noted observers of the present day—Koch, Pasteur, and Ehrlich—affirm that they have never detected bacteria in a healthy animal. Numerous attempts have been made to decide the question experimentally—by the history of healthy tissues, transferred under precautions against contamination, from the living or dying body to conditions of perfect isolation from bacteria. Such experiments demonstrate that some healthy tissues, at least, contain no organisms capable of inducing putrefaction; as the majority of bacteria are, however, incapable of effecting this process, the failure to putrefy does not necessarily prove the absence of all bacteria. Observation and experiment on the living body would also prove the absence of bacteria from healthy animals. The familiar fact that a dead human fœtus may remain in the mother's body for months or years without putrefaction, as in extra-uterine pregnancy, supports the same conclusion. Indeed, it has been repeatedly demonstrated that certain bacterial species, even when injected in considerable numbers directly into the blood or tissues of the living animal, cannot be found after the lapse of some hours; they appear to suffer the same fate as unorganized particles. Hiller even injected into his own skin some bacteria obtained from putrid flesh, and observed only a slight, transient, local œdema.

This failure of putrefactive and other bacteria to reproduce in healthy tissue seems to indicate their inability to maintain the struggle for existence against the animal cells indigenous to the soil. For seventy years a man may eat, drink, and breathe the ordinary bacteria, and carry a vast and varied assortment of them in his alimentary canal, without suffering putrefaction; yet so soon as his component cells are destroyed, generally, as in the death of the animal, or locally, as in the gangrene of a toe, the tissues swarm with these minute organisms. While some bacteria seem capable of development in tissues only after the death of their competitors, the animal cells, others exhibit this power in the presence of these cells when the tide is turned against the latter by the impairment of their nutrition, or by the presence in the blood of material favoring the invaders. The exudate on the cardiac valves frequently contains, not only in ulcerous, but also in simple "rheumatic" endocarditis, colonies of growing micrococci, a fact first observed by Klebs and confirmed by Eberth, Ehrlich, and Osler. In some cases it is demonstrable that the appearance of the bacteria upon these vegetations was subsequent to the inflammatory process. Abnormal blood composition seems to favor the development of some bacteria which may gain access to the tissues. The presence of glucose and increased metamorphosis of albuminous blood-constituents in diabetes are usually considered responsible for the proneness of diabetics to abscesses, carbuncles, cataracts, gangrene of wounds, sloughing of stumps even under most careful antiseptic dressings. These spontaneous abscesses and carbuncles contain, when freshly opened, micrococci in a high state of activity, as shown by Kraske, Eberth, and Pasteur. A clinical observation in the history of diabetics, their susceptibility to consumption, acquires peculiar significance in this connection, since the presence of the bacillus tuberculosus is now certainly a recognized anatomical characteristic, whatever may be said of its etiological relation. Cheyne refers to a weak, cachectic patient who had an abscess wherever he received a bruise, and whose abscesses always contain micrococci. Another question, formerly much disputed, is the possibility of invasion of animal issues by bacteria without previous solution of continuity. The only possibility for discussion now remaining is as to what shall be considered a solution of continuity; for it is definitely demonstrated that these organisms gain access to the body without the existence of any wound or abrasion discoverable upon the closest scrutiny. Trichinæ force their way from the intestine into the muscles: particles of coal are conveyed, probably through the agency of white corpuscles, into the parenchyma of the lungs, bronchial glands, and even the liver; it would seem *à priori* certain that many bacteria, much smaller than these particles, could also be received into the tissues. Ogston, using the most improved technique, found micrococci in every one of seventy previously unopened acute abscesses. Cheyne reports a similar experience. Micrococci have been repeatedly observed in the blood in spontaneous pyæmia, osteomyelitis, etc., and Obermeier's spirilla in recurrent fever patients, in cases where no lesion of the integuments was discoverable. But the question seems to have been finally decided by experimental demonstration: Buchner has induced anthrax in animals by the inhalation of powdered material containing the spores of the bacilli. When we remember the endothelial nature of the cells lining the pulmonary alveoli, we readily appreciate the facility with which particles, large or small, organized or unorganized, gain access to the circulation through the lungs.

Bacteria then, which, by virtue of their ubiquity, are in constant and frequently recurring contact with the animal body, are, like other minute bodies, organized and unorganized, frequently introduced into the body through solutions of continuity of the integuments, or through intact skin and mucous membranes, particularly by way of the lungs.

The burning question in pathology to-day is, in what degree are the various species of bacteria, present in human tissues during certain morbid conditions, to be regarded as the cause of the morbid processes with which they are respectively associated? Having already reviewed the conditions of eligibility for witnesses and of authenticity for evidence, we are prepared to consider in detail the present state of the question as to the individual diseases.

Before admitting the casual relation of a bacterium to the disease, we must be convinced not only that all observed phenomena can be easily reconciled with such assumption, but also that they can be as plausibly explained by no other assumption. The evidence of such causal relation must establish, therefore, 1, the competence of the observer and the accuracy of the observation; 2, the presence of a constant bacterial form in every case of the disease, and in numbers sufficient to explain the morbid phenomena; 3, the demonstrable isolation of the bacteria by successive cultures; 4, the induction of the disease in numerous healthy animals by inoculation with the isolated organisms; 5, the reproduction of the same bacterial form in the inoculated animal. Gauged by this standard, the evidence already adduced warrants the following unscientific but convenient classification:

*First.*—Disease, the demonstration of whose bacterial origin has been completed, through inoculation with isolated bacteria, by several competent observers—anthrax.

*Second.*—Disease whose bacterial origin has been affirmed, after inoculation with isolated bacteria, by one competent observer—tuberculosis.

*Third.*—Diseases which are uniformly characterized, intra vitam, by the presence of bacteria in the tissues; but which have not as yet been induced by inoculation with the isolated bacteria—recurrent fever, pyæmia, diphtheria, erysipelas, leprosy, rhinoscleroma, gonorrhœa (urethræ et conjunctivæ), and some forms of septicæmia. Puerperal fever, osteomyelitis, and ulcerous endocarditis are considered to belong etiologically with pyæmia, septicæmia, erysipelas, and diphtheria.

*Fourth.*—Disease after death from which (in some cases also during life) bacteria have been observed in the tissues: variola, scarlatina, typhoid fever, croupous pneumonia.

*Fifth.*—Diseases in which the presence of bacteria, ante or post mortem, has been asserted: syphilis, intermittent fever, yellow fever, typhus, measles, lupus, rabies, tetanus, *et al.*

The investigation of the diseases included in the fourth and fifth classes is as yet quite imperfect and inaccurate, partly at least because they are with few exceptions peculiar to man.

The evidence as to the relations of bacteria to disease rests largely upon experimental observations upon the lower animals. The impossibility of thorough examination of human tissues, not only during life, but also for hours or days post-mortem, restricts materially the field of clinical observation on this subject; for a few days, even hours, suffice to people a dead body with bacteria. We may simply ignore such contributions as that of Zander (*Virch. Arch.*, Bd. LIX.), in which he announced the probable discovery of the bacterial origin of acute yellow atrophy of the liver. In one fatal case of the disease he found bacteria in the liver, the section having been made fifty-four hours after death. He anticipates the possible objection that the bacterial development was perhaps post-mortal, and refutes it with the convincing statement that "the body did not, so to speak, exhibit any marked symptoms of putrefaction." Indeed, while defective methods may be chiefly employed by the practically inexperienced, most egregious misinterpretation is unfortunately not monopolized by them. Several pioneers in this subject, notably Pasteur and Klebs, have, by the publication of hasty and, as subsequently appeared, erroneous conclusions, forfeited much of the prestige acquired by their earlier classical works on fermentation and gunshot wounds respectively. These two observers have ignored successive improvements in technique; have apparently assumed that all infectious diseases are of bacterial origin; that therefore the discovery of a bacterium in a diseased animal is ample proof of its pathogenetic influence. Years ago, Klebs announced the discovery of the bacterium of tuberculosis, and even declared that the disease could be cured by an agent, benzoate of sodium, which destroys the parasite. With equal facility he discovered the bacterial origin of syphilis, typhoid fever, etc.

Pasteur seems to have a fondness for micrococci, especially in the figure of 8 form, to which he ascribed puerperal fever, rabies, his "nouvelle maladie," as well as chicken cholera, and typhoid fever. His famous publications about the inoculation of chickens with anthrax, and about the rôle of earthworms in transporting anthrax spores, further illustrate the inability of his judgment to cope with his imagination.

## Lecture II.

### SEPSIS AND ANTISEPTIC SURGERY.

The infectious diseases usually consequent upon wounds, and therefore falling within the province of the surgeon, merit somewhat extended discussion. Such diseases were known to the earliest medical writers, and, we may assume, antedate man himself by so much time as the existence of highly organized life preceded him. Clinical experience had, long before the advent of experimental research, identified infectious septic material with the products of that complicated process whereby lifeless organic substances are de- and re-composed into chemically less complex matters — putrefaction; and early experimental investigation was directed to the elucidation of that process and its relations to disease. Gaspard first proved experimentally that the injection of putrefying substances of animal or vegetable origin—blood, pus, bile—was followed by the clinical features of sepsis. Panum demonstrated that the putrid infectious substance is not gaseous; that it is not destroyed by eleven hours' boiling and complete desiccation; that it is insoluble in alcohol, but present in the watery extract of putrid materials, even when dried; that the albuminous matters in putrid fluids are not *per se* septic, but condense the infectious matter upon their surfaces; for the filtrate, containing no solid particles, preserves the septic properties unimpaired. Panum concluded that the putrid agent must be a definite chemical compound like curare and the alkaloids, and named this hypothetical substance sepsin. Billroth, Weber, Hemmer, and Schweninger repeated and confirmed these experiments. The last-named observer concluded from the fact that different results followed the administration of the same quantity of the same putrid liquid at different stages of putrefaction, that not one only, but various putrid products, arising at different periods of the process, possess septic properties. Bergmann and Schmiedeberg obtained from putrefying beer-yeast a crystalline substance, melting on contact with air and charring under heat, which induced in dogs the clinical and anatomical appearances of sepsis. This they termed sulphate of sepsin. Zuelzer and Sonnenschein isolated a similar compound. Hiller and Mikulicz demonstrated that the septic agent of putrid materials could be extracted and retained by glycerine, and in so far was analogous to the active ingredient of vaccine lymph, to pepsin, ptyalin, etc.

It was thus established, and so remains, that the clinical and anatomical features of septicæmia could be induced by unorganized substances obtained from the products of putrefaction. Yet in these cases two characteristics frequently observed in the septic infection of human subjects were often conspicuously absent—the stage of incubation and the infectiousness of the septic blood and tissues. Panum noted particularly that the influence of his boiled putrid materials became manifest in fifteen minutes to two hours, and attained its acme in four to eight hours. Meanwhile a new path of investigation had been opened by Pasteur's demonstration that the putrefaction of animal tissues is a phenomenon incident to the vital activity of certain bacteria—facts established incontestably by the researches of Pasteur, Tyndall, Traube, Brefeld, and their pupils. The determination of the relation between the bacteria and the diseases caused by the putrid products of their vital action soon became the object of most patient and careful investigation. Coze and Feltz found vibrios intra vitam in the blood of animals infected with putrid fluids; and similar organisms post-mortem in the blood of a patient dead of putrid infection. With this blood they inoculated a rabbit, which then exhibited septic symptoms, and whose blood was found to contain similar vibrios. Rindfleisch found colonies of bacteria in the heart-muscle from a case of pyæmia; Recklinghausen and Waldeyer followed with similar discoveries. The work of Klebs on "Gunshot Wounds" (1872) opened the new epoch in pathological investigation. The examination of numerous gunshot wounds, both before and after death, showed that the organs and tissues exhibiting morbid changes due to such wounds were populated by bacteria; serous surfaces, both those opened by the bullet and those which, though still intact, lay adjacent to an abscess or to the track of the missile; the walls of blood-vessels, not only those which had been the seat of secondary hemorrhage, but also those which, while not ruptured, showed beginning thrombus formation; metastatic abscesses in liver and lung; leucocytes in and near the track of the bullet—all contained colonies of bacteria. A series of experiments upon animals showed that while the injection of putrid liquids, containing naturally myriads of bacteria, was followed by continuous fever and metastatic abscesses—*i.e.*, pyæmia—the injection of the same liquids after filtering through clay and thus deprived of solid particles, including the organisms, was followed by fever just as intense, though transient, but never by metastatic abscesses—*i.e.*, septicæmia.

The work of Klebs, which proved that there must be some intimate relation between the pathological processes and the bacteria, was soon followed by a series of accurate experimental observations by Samuel, from which he concluded that the varying effects of putrid fluids upon the living animal were due to various substances therein contained; that the specific septic (toxic) influence is the effect of certain volatile matters, probably combinations of sulphur and of ammonia; to the bacteria he ascribed the influence whereby the infection is localized progressively in various organs remote from the original wound.

Billroth concluded, as the result of much careful clinical and experimental observation, that the presence of bacteria was the result and not the cause of certain changes in secretions and tissues. He assumed the formation, during inflammation and putrefaction, of a "zymoid" substance whose presence (1) conferred upon pus and exudate their infectious character and (2) converted wound secretions into favorable soil for organisms. He believed, however, that bacteria might be the means for transporting and multiplying his hypothetical zymoid, and in this capacity might be, probably are, the carriers and originators of specific pathological processes. Since the publication of this work (1874), Billroth has materially modified certain of his conclusions.

From these investigations it was generally concluded that septic infection was due to an unorganized though perhaps organic substance; that the presence of bacteria was an epiphenomenon—a sequence, not a cause; that their deleterious agency, if any, consisted simply in the transfer of unorganized infectious matters from one part of the body to other portions, perhaps from one individual to another.

But there soon appeared from various sources, notably Koch and Pasteur, investigations more or less incompatible with these views. Pasteur found that in the serous sacs, muscles, liver, and spleen of a septically infected animal there are always present microscopic organisms (*microbe septique*), although the blood may be until death free from them. Inoculation with a drop of peritoneal serum, or a piece of muscle from an animal dead of sepsis, induces, in a second animal, all the appearances, ante- and post-mortem, of the original disease; while a drop of blood from the heart-cavity (proven microscopically to contain no septic vibrios) is, on the contrary, innocuous. Pasteur cultivated his vibrio septique in various fluids, such as solution of beef extract, in the manner already described; and found that a drop of fluid from the last flask, containing presumably none of the original unorganized septic matters, but crowded with the vibrios, produced the original septic disease. In the tissues of infected animals Pasteur was unable to find any unorganized substance capable of inducing sepsis, as had been affirmed by Panum, but he found that a putrid fluid, a few drops of which induced sepsis and death, lost its poisonous properties entirely in a few hours, when exposed in a thin layer to the air. Now, since the vibrios likewise lose their vitality in a few hours in the presence of free oxygen, Pasteur insists that the loss of virulence in the fluid is due solely to the enforced inactivity of the contained vibrios. Yet he is inclined to the belief that just as the alcoholic fermentation of grape sugar is a vital phenomenon manifested by any one of several species of fungi, so the production of septic substances may accompany the vital activity of any one of several different bacteria.

Pasteur further reports that among the organisms usually present in ordinary water is one identical morphologically with the bacterium termo, but physiologically distinguished by the fact that its injection from an isolated cultivation under the skin of a rabbit is followed by abscess formation at the site of puncture. Injection of the same organisms directly into the circulation, or in several places subcutaneously, is followed by the formation of abscesses in lungs and liver; by fever and death—in short, by pyæmia. A piece of the liver or lung develops in a culture liquid, the same micrococcus in great numbers. Such liquid, if previously boiled however, so as to destroy the organisms therein contained, causes, upon subcutaneous injection, abscesses as before, but without general infection of the animal. He rejects, therefore, for pyæmia as for septicæmia, the agency of an unorganized, soluble septic agent, and considers the bacteria alone responsible for septic infection.

Markedly different was the reception accorded to a monograph published in 1878, by a then almost unknown young German physician, Robert Koch. He surmounted, by improvements in technique, some of the hitherto insuperable difficulties in the recognition and investigation of bacteria—improvements which confer such evident and extreme advantages that they have become absolute necessities for original research in this field; indeed so many errors of commission, as well as of omission, have been thereby detected in other methods, that one is disposed to regard as uncertain any researches in which these measures most essential to accuracy of observation are neglected. Having demonstrated the life history of the bacillus anthracis, which the French school, working with Pasteur's clumsy method, had for sixteen years failed to discover, Koch turned his attention to the etiology of surgical infectious diseases. He found that the subcutaneous injection into a mouse of five drops of putrid blood was followed by immediate prostration, and in four to eight hours by the death of the animal. There occurred in these cases no local reaction, the internal organs were apparently normal, no bacteria were detected in the blood or tissues, inoculation of other animals with the blood from the heart caused no perceptible effect. Koch considers this disease therefore as septicæmia, etymologically as well as clinically—the introduction into the blood of a poisonous substance, soluble, not reproducing itself, analogous, in fact, with the effect of certain vegetable alkaloids and of ptomain, the substance isolated by Selmi from human corpses, which so closely resembles atropine in its physiological effects. This is also the effect obtained from the injection of boiled putrid materials by Panum, Bergmann and Schmiedeberg, and others. Koch found, however, that the injection of a smaller quantity, one-half to one drop of the same putrid blood, was followed by entirely different effects. In some cases the mouse was apparently unaffected; in others brief, transient depression was observed; in perhaps one-third of the cases there ensued, twenty-four hours later, progressive weakness, retardation of respiration, drowsiness, and, in forty to sixty hours, death. Section revealed no other pathological changes than local œdema at site of inoculation and decided enlargement of the spleen; but after inoculation of a second mouse with a minute quantity (one-tenth to one-half drop) of liquid from this œdema, or of blood from the heart, the latter animal presented, in forty to sixty hours, precisely the same clinical and pathological picture as the first; from the second, a third was successfully inoculated, and so on *ad libitum;* indeed, the mere contact with a fresh wound of a scalpel-point previously dipped in the septic blood sufficed. Here, then, was something entirely different from the intoxication following injection of a larger quantity (five to ten drops)—differing in the existence of a marked stage of incubation, of local reaction, and in certain and uniform infectiousness. The blood of such an animal evidently contained something not present in that of the former mouse—a something requiring time for the manifestation of its influence, and finally distributing itself throughout the entire blood-mass, so that each drop thereof possessed the septic possibilities of the original putrid drop. Such mode of action implies reproduction, and reproduction is a characteristic of organized matter. It was to be expected *à priori*, therefore, that the blood contained organisms; Koch found, in fact, invariably, that the blood serum, white blood-corpuscles, and various tissues of such animals swarmed with minute rods, which stained readily with aniline colors, and when removed from the body into similar artificial conditions multiplied by transverse fission. Since the blood of the infected and infecting mouse differs evidently from that of the intoxicated and non-infecting mouse only in the presence of these bacteria, Koch ascribes the infectiousness to these organisms. It is interesting to note that all attempts to inoculate rabbits and field-mice with the septic blood were fruitless. The animals remained unaffected; no bacilli were found in their blood, although the mouse-blood used for inoculation was full of them. Further, that although the putrid fluids injected contained organisms of numerous varieties—micrococci, bacteria, bacilli—all of which were subsequently found in the local œdematous liquid, yet only one species, the minute bacillus, was found distributed throughout the blood and tissues. The living mouse seemed to be a culture-medium for isolating these from the other varieties, to whose growth the animal's tissues were less perfectly adapted. In the second or third mouse successively inoculated only the specific bacilli were found.

By methods essentially similar, Koch demonstrated the association of another form of septic infection of mice, with a micrococcus species; of septicæmia in rabbits with a bacterium; and of pyæmia (with metastatic abscesses) in the latter animal with a micrococcus variety. Yet, although the fact of association was amply demonstrated, there still remained the possibility of objection that the essential agent in the infection was a soluble, unorganized substance contained in the putrid liquids and the infectious tissues used for inoculation, the bacteria being the result and not the cause. The final demonstration to the contrary, the proof that these different effects could be induced by inoculation with the respective bacterial varieties after complete isolation from accompanying animal tissues by cultivation upon solids, was not furnished at the time of this publication by Koch, but has been subsequently completely established by Koch, Gaftky, and Löffler in the laboratory of the German Health Bureau, for the bacilli of mouse- and the bacteria of rabbit-septicæmia. The extreme accuracy and critical supervision of manipulations, the logical sequence of methods, and withal the unpartisan candor and earnest desire for truth evident throughout this work of Koch's, inspired at once a confidence which has not as yet been diminished nor betrayed. It is worthy of note that the infectious disease of mice described by Koch as malignant œdema is identical in clinical and pathological appearances with that which Pasteur ascribes to his "vibrio septique;" while the pyæmia of rabbits corresponds accurately with the purulent infection which, according to Pasteur, follows the injection of his microbe of pus. These results, obtained quite independently by two observers, using different methods, have been confirmed not only by Gaffky and Löffler, but also by Rosenberger in Würzburg, in a series of carefully performed experiments.

A review of the evidence already considered shows, then, that infectious diseases, identical in clinical and anatomical appearances with the various forms denominated septicæmia in man, have been induced in the mouse and rabbit by inoculation with animal tissues in various stages of putrefaction; that the resulting infection is just as certain if the putrid substances be previously boiled and thereby deprived of living organisms (Panum, Bergmann, Rosenberger). On the other hand, it is certain that *per se* innocuous culture fluids—infusions of beef, etc.—acquire, after inoculation with minute quantities of infected blood or tissue, the same septic properties, provided such blood or tissue contain living bacteria; it is further certain that this multiplication of the septic substance in such liquid is a concomitant of the vital action of the organisms therein contained (Pasteur, Koch, Rosenberger); it is further demonstrated that these organisms can and do, not alone multiply the septic material, but when isolated by successive cultures from all the accompanying animal tissues, induce, independently, fatal infectious disease (Pasteur, Koch, Löffler, Gaffky, Rosenberger).

The same principle—vital activity of bacteria—pervades all these phenomena; for the artificial induction of septic diseases has been, in all these experiments, originally accomplished by the incorporation into the animal of *putrid tissues*, with or without bacteria. Now, since putrefaction must be regarded, in the present state of our knowledge, as impossible without the presence of these organisms, it is evident that sepsis, putrid infection, was in every case due, directly or indirectly, to the action of bacteria; since even the boiled substances used by Panum and Rosenberger, and the sepsin obtained from rotten yeast by Bergmann and Schmiedeberg, had acquired their septic properties through putrefaction, *i.e.*, through the action of bacteria. Hence we are logically driven, by all this work, to the belief that septicæmia implies the introduction into the animal either of living bacteria, or of a substance which has acquired noxious properties through previous vital activity of these organisms.

More recent experiments have demonstrated, however, that the etiology of the group of clinical and anatomical appearances known as septicæmia is by no means restricted to putrid infection. In the researches as to the nature of blood coagulation, instituted by Schmidt, of Dorpat, and his pupils, it was noticed incidentally that the introduction or production in the blood of fibrin-ferment in considerable quantity produces effects identical with those of putrid infection—septicæmia. In this case the result is of course attributable to coagulation of the blood. Similar phenomena were observed by Köhler, Angerer, Naunyn, and Francken, to follow intra-venous injection of fresh blood-serum (containing therefore both fibrinoplastin and ferment); of hæmoglobin solution (which is known to favor the formation of fibrin-ferment in the blood); of sulphuric ether (which sets free hæmoglobin and hence indirectly fibrin-ferment). Injected in large quantities, these substances caused immediate death by instant coagulation of the blood in the heart and large arteries; after smaller quantities the animals survived hours or days, and exhibited the usual symptoms of septicæmia; their blood contained free fibrin-ferment, while that of healthy animals does not. Finally Edelberg, working under Schmidt's direction, established clearly that the injection of fibrin-ferment alone, isolated from other ingredients of blood, can induce the same phenomena.

In a series of experiments communicated to the Congress of German Surgeons, in 1882, Bergmann observed the clinical and anatomical features of septicæmia—fever, swelling of spleen and lymph-glands, gastro-intestinal inflammation, cardiac weakness, ecchymoses in mucous and serous membranes—after the injection of the physiological ferments, pepsin and trypsin, in small doses; large quantities induced, like fibrin-ferment, immediate death by coagulation of blood in the larger vessels.

Raynaud and Lannelongue inoculated rabbits with saliva from a child dead of rabies, and induced thereby an infectious disease, terminating fatally in forty-eight hours or less. Pasteur found in the blood of these animals a bacterium which he regarded as the cause of the disease.

Inoculation of rabbits with saliva from children dead of broncho-pneumonia caused the same result, and produced the same figure-of-8 bacterium. The same organism was found in the saliva of a healthy adult. Sternberg found that injection of fresh saliva from certain healthy individuals caused a similar fatal infectious disease, which he calls septicæmia, in rabbits, characterized by the presence of a micrococcus apparently identical with Pasteur's; and asserts that this organism, isolated by flask cultures, induced the disease again upon subcutaneous inoculation. Neucki and Gautier isolated from saliva a substance capable also of producing fatal infection of certain animals. Saliva, then, not only after death of the subject, but even fresh from the living individual, can also induce septicæmia. Whether the effect shall be ascribed to a contained bacterium or not is immaterial to our present purpose, which is to emphasize the fact that the group of phenomena called in general septicæmia may follow other causes than putrid infection; may be induced on the one hand by the vital action of isolated bacteria, and on the other by unorganized substances—the boiled septic materials of Panum and Rosenberg, the sepsin of Bergmann, the fibrin-ferment of Edelberg, pepsin and trypsin of Bergmann, hæmoglobin, etc.

The mode of action common to several, at least, appears to be the liberation of fibrin ferment; for the blood of septicæmic animals is characterized by the presence of free ferment, which is not found, unless perhaps as traces, in normal blood. This ferment seems to arise, according to the researches of Schmidt, in the disintegration of white blood-corpuscles; and these are known to be invaded and apparently disintegrated by bacteria, in the septicæmia of mice and rabbits, at least. It would appear, although not for all cases demonstrated, that the clinical and anatomical features common to the various forms of septicæmia are attributable to the rapid liberation of fibrin ferment in the blood; and that any agent —organized or unorganized, putrid or fresh—capable of effecting such liberation may induce the disease.

This conception, at any rate, enables us to understand much that is otherwise perplexing. Various have been the attempts, for example, to explain the so-called aseptic wound fever, which occurs in the majority of severer wounds, even under the most perfect Lister dressing. Küster and Sonnenburg ascribed it to absorption of carbolic acid ; but extensive experiments upon man, as well as the lower animals, have proven that the acid does *not* cause fever ; but induces, on the contrary, after slight, brief, and by no means constant elevation, a decided depression of temperature. Others have referred the phenomenon to absorption of chloroform—a hypothesis incompatible with the fact that wound fever follows operations performed without anæsthesia (as is so often done in Germany and Austria) as usually as those done under chloroform. The more general opinion, that aseptic wound fever differs from sepsis,—*i.e.*, putrid infection—in degree rather than in kind, meets a serious obfection, as Gussenbauer remarks, in the fact that the jormer occurs within a few hours after the infliction of the wound, before decomposition and consequent sepsis can be reasonably presumed to have occurred.

The clinical facts—(1) that a large minority of wounds, severe as well as slight, are followed by no fever under the Lister dressing, as was the case in over three hundred of a thousand reported by Volkmann and Genzmer, and in nine of twenty-four most carefully observed by Edelberg ; (2) that the course of subcutaneous fractures without extravasation of blood is usually afebrile, while similar fractures with extensive blood extravasation often induce fever ; (3) that the application of a tight bandage to a wound or fracture, which must cause some extravasation of blood, is often followed by fever in a patient previously afebrile (Edelberg) ; (4) that the blood of patients during simple surgical fever sometimes contains free fibrin ferment in appreciable quantities—such facts indicate that aseptic wound fever is caused by absorption from extravasated blood, especially since it has been demonstrated, as already remarked, that blood, fluid or coagulated, hæmoglobin, or even isolated fibrin ferment can experimentally induce the same phenomena. It is further conceivable, though not demonstrated, that the products of a local inflammation, or the modification of cell-activity through fatigue or emotion, may also be directly responsible, through destruction of leucocytes and liberation of fibrin ferment, for some of those cases of spontaneous septicæmia which we ordinarily ascribe to unperceived entrance of bacteria or putrid products into the body.

Septicæmia is, then, a collective name for processes more or less similar, but etiologically distinct—at least, in certain lower animals; any one of several unorganized substances, any one of several bacteria (at least, three in the case of the mouse) may induce characteristic symptoms. It has been proposed to adopt the term *sapræmia* for putrid infection without bacteria, retaining the usual name to indicate the effect of organized agents; yet the clinical distinction is probably rarely possible.

The clinical experience of all ages has unanimously ascribed the second type of septic infection—characterized by chills, a remittent or intermittent fever and the formation of multiple abscesses—to absorption from pus; and it has always been designated by a name--pyæmia, purulent infection—indicative of this supposed origin. The discussion of the relations of bacteria to pyæmia begins, therefore, naturally with the consideration of their relations to suppuration. That these organisms should exist in pus exposed to the air, as in other albuminous liquids under like conditions, was *à priori* probable and long ago demonstrated; that they exist also in the pus of abscesses which have never been opened, has been conclusively demonstrated by Klebs, Nepveu, Rindfleisch, Waldeyer, Cheyne, Ehrlich, and especially by Ogston, who found micrococci in every one of seventy previously unopened acute abscesses, though rarely in chronic, cold abscesses.

The mere fact of association does not, of course, necessarily prove a causal relation of the organisms to the suppurative process; but the observation that a zoöglœa mass of micrococci is often the centre of an abscess; that indeed abscess-formation in all stages, from a simple accumulation of straggling leucocytes to the fully developed destructive infiltration of tissue, has been observed

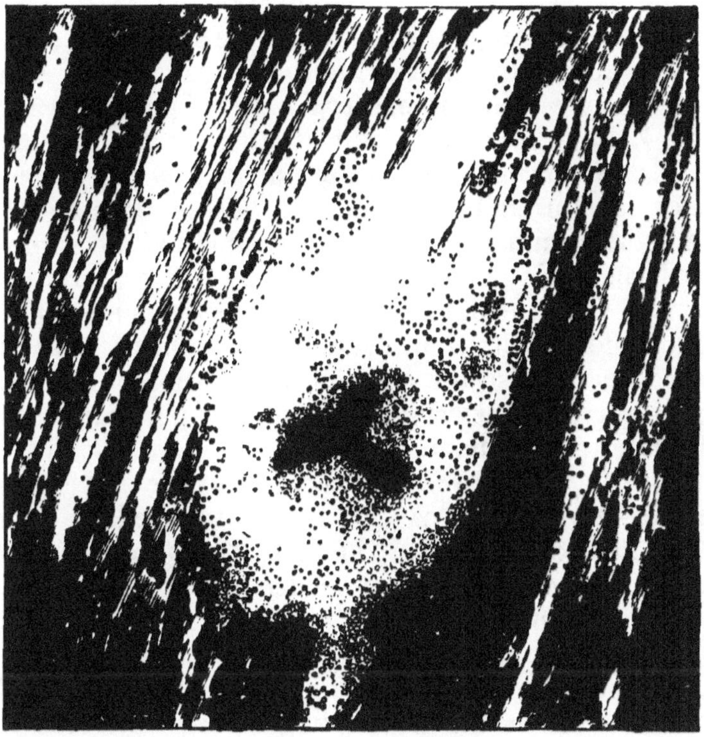

Fig. 6.—Incipient abscess formation around an embolus of micrococci; heart muscle in a case of endocarditis ulcerosa. × 100. (Copied from a photogram by Koch.)

around a nucleus consisting of a minute embolus composed entirely of micrococci (Fig. 6); that the progressive purulent infiltration of the surrounding tissue is preceded by an invasion of the same by micrococci—can be impartially and satisfactorily explained, in the present state of knowledge, by no other hypothesis than that the micrococci *cause* the suppuration. Experimentally there is direct evidence to the same effect.

Pasteur saw, after cultivation of a micrococcus found in ordinary water, that the injection of a few drops of the previously harmless culture-fluid, now containing myriads of micrococci, was invariably followed, in the rabbit, by suppuration around the point of injection, the pus and tissues containing numbers of the same organisms. The intra-jugular injection of the same fluid caused multiple abscesses in the internal organs. He found the same micrococcus in pus from cases of puerperal fever. Klebs, Zahn, and Tiegel found that while the injection of pus from a pyæmic abscess or putrid fluid was followed by local suppuration and multiple abscess formation in the infected animal, the same pus or liquid, after filtration through clay cylinders—whereby the bacteria were separated from the liquid—caused intense general infection, but no suppuration, even at the point of injection. Koch observed also the constant association of a characteristic micrococcus with infectious suppuration in the rabbit after putrid inoculation.

It appears, therefore, impossible to evade the conclusion that suppuration can be and is induced by micrococci. That this effect is induced by one or more specific varieties of these organisms seems probable from these researches of Klebs, Koch, and Pasteur; that it is not induced by *all* species is apparent from the fact that colonies of micrococci are frequently present in the human and other animals during various morbid processes in which suppuration does not occur—as in erysipelas. As to the mode in which this influence is exerted, there is no definite knowledge; the assumption that the deleterious effect results from changes in the chemical constitution of the containing medium, as an essential feature of the vital activity of these organisms, is supported by analogy with the processes of fermentation and putrefaction, by the phenomena known to attend the life of other bacteria, and by the direct observations of Koch and Pasteur.

Yet the induction of suppuration is not a monopoly of micrococci: the growth of the actinomyces bovis in the tissues is accompanied by the formation of abscesses; upon microscopic section the fungus is found constituting the nucleus of a miliary abscess; and inoculation with the isolated actinomycetes proves that the fungus itself, and not a hypothetical soluble substance accompanying it, is responsible for the suppuration. Experimental researches upon suppurative keratitis by Leber, of Göttingen, make it highly probable that another bac-

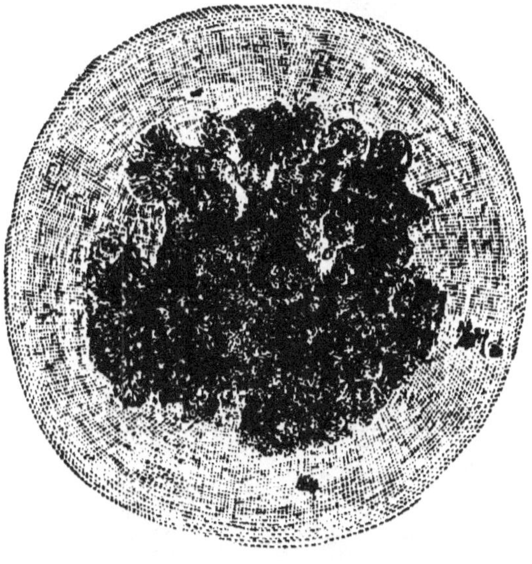

Fig. 7.—Actinomyces bovis—tongue of ox. × 140.

terium, the leptothrix of the mouth, and a mould fungus, aspergillus glaucus, can also induce suppuration. On the other hand, several bacterial varieties are known to inhabit at times human and other animals without causing suppuration—the bacillus anthracis, for example.

The school of pathologists of which Weigert is a prominent exponent, has been inclined to regard micro-organisms not merely as a cause, but as the sole cause of acute suppuration. This view was certainly incompatible with many clinical observations, and has been recently quite dissipated by experimental research. Uskoff, under Ponfick's direction, has shown that the subcutaneous injection into rabbits of turpentine oil, is followed by the formation of pus in which no bacteria can be detected ; an assertion which has since been corroborated by others. The most accurate and conclusive

Fig. 8.—Actinomyces bovis, and pus corpuscles. × 260.

experiments in this direction were recently made by Dr. Wm. T. Councilman, of Baltimore, in the laboratory and under the direction of Cohnheim. Councilman made a number of glass capsules, heated them to redness, poured into each, while still hot, a boiling mixture of croton and olive oil, and sealed the open end in the flame. Fourteen of these were inserted, with antiseptic precautions, under the skin, at various places, in different rabbits. In no instance was suppuration or even adhesive inflammation observed—the capsules remained freely movable in the subcutaneous tissue. After intervals varying from two to fourteen days, when the incision wound was firmly healed, the capsules were broken subcutaneously, by force applied to the skin; in every instance suppuration followed in a few hours. An examination of the pus and abscess walls revealed nothing that could be recognized as bacteria.

Chronic or cold abscesses may have a different etiology—their clinical history and appearances, and the fact that they seldom cause pyæmia, would point to that conclusion. It is significant that of eighteen chronic abscesses, Ogston—employing the most approved technique—could detect bacteria in but four, which were consequent, moreover, upon erysipelas, typhoid fever, pharyngitis, and pulmonary consumption, respectively; in the remaining fourteen, both microscope and attempts at cultivation gave only negative results. Their absence at the time of examination does not exclude the possibility of their presence at an earlier period of the process; that they may have lost their vitality, and with it their power to absorb the coloring agents.

Suppuration must be regarded, then, as indicating the presence of an element foreign to the living animal cells; which may be introduced directly, like the croton oil in Councilman's experiment, or indirectly as an incident in the life of various fungi. That a derangement of cell-nutrition, local gangrene, may by mechanical or chemical irritation, without the presence of other organisms, effect the same result, seems probable in view of clinical experience, but is not yet experimentally proven. Practically, we may regard acute suppuration as proof of the access of external irritant matter, organized or unorganized; and clinically, we must agree with Cohnheim, that suppuration not due to bacteria or other fungi is extremely rare. The comparative rarity of pus-formation under the Lister dressing—although this is, at best, an uncertain means of excluding organisms—is highly significant of the relations between the two.

That form of septic infection known as pyæmia is distinguished by abscess formation in external organs—from which fact alone it is evident that bacteria must play an essential part in the disease. Yet there is abundant direct evidence to the same effect: these metastatic abscesses always contain bacteria; these organisms exist not simply in the pus and in the inflamed tissue constituting the abscess wall, but also in a zone external to the territory already involved in inflammation; the invasion by organisms may therefore apparently precede the inflammatory re-action; further, incipient inflammation and suppuration are observed around minute emboli consisting of micrococci; and finally, although thrombosis and embolism occur in various pathological conditions, no suppuration occurs in such fibrinous masses, nor in the adjacent tissues, unless bacteria also be present. These anatomical facts are quite in accord with the independent evidence of experimental research. Panum found that the injection into the jugular vein of minute balls of wax or mercury caused the formation of emboli in the lungs, but that no suppuration occurred around them; he then combined embolic formation with putrid infection by the injection of putrid fluids just before or just after the formation of emboli, through the incorporation of wax, mercury fibrin or cheese particles; and varied the experiment by the artificial induction of superficial phlegmonous inflammation, in the course of which the formation of emboli was secured by injections as before. The result was always the production of simple, non-suppurating embolic masses; the substitution of fresh blood-clots for the wax, mercury, etc., whereby a closer approximation to the natural embolic process was secured, gave the same results. Emboli caused by intra-venous injection of particles of putrid flesh, however, were promptly followed by suppuration. Other experimental researches into the formation of embolic (metastatic) abscesses—by Virchow, Billroth, Weber, Waldeyer, Cohnheim—

confirmed these results of Panum; putrid emboli always softened and excited suppuration; others rarely; it was further established (Waldeyer) that puriform softening of a thrombus can be caused by contact of pus or putrid matters with the external surface of the containing vessel, as well as by admission to its lumen. Since suppuration in the immediate vicinity of a vein may cause inflammation and thrombus formation in the vessel, it is apparent that phlebitis, puriform softening of thrombi—in short, pyæmia—may occur without any artificial solution of continuity in the vascular walls. Experiment has always shown that fluids (pus and putrid matter) capable of inducing pyæmia, lose by boiling (Panum, Bergmann, Pasteur) or by filtration (Klebs, Zahn, Tiegel) this power to cause metastatic suppuration—pyæmia—though still able to induce rapid and fatal infection—septicæmia. Since by these measures—boiling and filtration—the contained organisms are destroyed or eliminated, experimenters are unanimous in ascribing the induction of metastatic abscesses to bacteria.

The clinical evidence is almost as strong; for, according to the unanimous assertions of eminent surgeons—Nussbaum, Volkmann, Esmarch, Thiersch, Verneuil, Schede, Gussenbauer, for example—pyæmia is practically unknown after wounds which have been treated from their inception by the Lister method, the avowed object and essential feature of which is the attempt to exclude organisms.

Perhaps the strongest clinical evidence of the septic influence of bacteria is afforded by the cases of so-called spontaneous pyæmia, where no suppuration nor solution of continuity is detected. In many of these a closer search would doubtless reveal a possible source of purulent infection. Weichselbaum has recently called attention to fatal cases of this kind in which the focus of infection was found as suppuration in the nose and antrum. Yet there still remain numerous cases in which pyæmia appears to proceed from deeply situated abscesses, which can have had no direct communication with the external world

—after subcutaneous fractures, for example; and still another class in which a general infection without local suppuration during the first few days occurs without exciting cause, unless perhaps exposure to cold; and until the appearance of pus in the joints, etc., cannot be distinguished from acute rheumatism or from other infectious diseases. To this category belong cases of acute osteomyelitis and ulcerous endocarditis. The blood and metastatic abscesses contain in these cases also the usual micrococci; the history presents, in fact, nothing unusual except the obscurity of the infection. In some of these—as in one of osteomyelitis reported by Gussenbauer—the bacteria were observed in the blood and in the bone-marrow before suppuration had occurred; the general infection preceded the local affection. Such cases must incline us decidedly to the view that the micrococci caused not only the local suppuration, but also the primary general infection. It is noteworthy that such cases of primary pyæmia often follow exposure to cold; perhaps we should regard the retention of certain material in the blood, this interference in excretion, as a predisposing moment which has favored the development of organisms; diabetic patients certainly are especially prone to local gangrene and septic infection after a wound, and it is equally well known that a minute incision, even needle-puncture of the dropsical skin in amyloid degeneration of the kidney, exposes the patient to erysipelas and pyæmia. Yet in some cases the bacteria essential to pyæmia can and do exhibit their vital activity in the human body without the pre-existence of any recognizable deviation from the usual health, and without any discoverable solution of continuity in the integuments.

In this discussion I have assumed the etiological identity of the septicæmia and pyæmia of man with that of processes marked by the same clinical and anatomical features in animals. To such assumption objection may be made, based on the known differences in the effects produced on man and other animals by the same toxic agent; rabbits, for instance, live and fatten on a diet of belladonna leaves, and carnivorous animals are but slightly susceptible to anthrax. But when we consider that the septic processes of man are objectively identical with those of other animals, that they result in various animals alike from putrid and purulent infection, and furthermore, that they have been induced in animals by direct inoculation from the human subject, we must justify the application to man of the principles ascertained from the study of these septic diseases in other animals.

Although our present knowledge of the etiology of septic infection is thus incomplete, our ability to prevent such infection is fortunately more satisfactory. For we may practically classify all such cases into two categories—those in which a possible source of infection is previously apparent, and those in which no such source is discoverable. The treatment of the former class, the large majority, comprising all wounds, I may be permitted to discuss in so far as the principles of such treatment are based upon a recognition of the agency of bacteria in the morbid processes. Septic infection from a wound means the absorption through that wound of one or more constituents of the putrefactive process. Now, putrefaction is impossible without bacteria; hence septic infection implies the vital activity of bacteria, past or present. The actual presence of bacteria in the wound is, as has been shown by Panum, Bergmann, and others, unnecessary—septicæmia may be induced by putrid liquids deprived of bacteria; *but these liquids are putrid*—they embody the products of bacterial life. Precaution against the introduction into a wound of already formed poisons—by disinfection of hands, instruments, sponges, etc., on the side of the surgeon, and by similar cleanliness as to the body of the patient, is evidently the first measure against sepsis—a measure quite overlooked occasionally by surgeons who intend to use all so-called antiseptic precautions. I once saw a laparotomy made by a rigid apostle of Listerism; the carbolic spray was used; hands, instruments, ligatures, etc., thoroughly carbolized; but the patient's skin was not even washed; several coils of intestine were in course of the operation laid upon the skin, and came in contact with the pubic hair. The operation itself was not formidable, but the woman died of purulent peritonitis.

If no infectious matter be thus carelessly introduced from without, the occurrence of sepsis from a wound necessarily implies decomposition in the wound itself. For the accomplishment of such decomposition it is evident that three factors must concur: 1, the presence of animal tissues deprived of vitality, and hence capable of putrefaction; 2, the presence of organisms capable of inducing putrefaction; 3, the prevalence of conditions which permit the vital activity of these organisms. The absence of any one of these conditions renders putrid infection impossible. We are familiar with analogous phenomena outside of the body. Urine or blood in free contact with ordinary air putrefies; if access of bacteria be prevented by closing the mouth of the test-tube with cotton, etc., putrefaction does not occur; the process can be prevented with equal certainty by changing the environment—addition of alcohol for example—whereby the vital activity of bacteria is arrested. We have abundant evidence, as has been already stated, that the same principles prevail within as well as without the living animal. That the bacteria ordinarily present in the air are powerless to destroy living tissues is proven by the fact that unfiltered ordinary air has been passed for hours through the peritoneal cavity of rabbits without inducing pathological changes—indeed, the entire subcutaneous tissue of animals has been inflated with air with like result; by the harmlessness of surgical emphysema; by Hiller's injection of such bacteria into his own body, etc. That the presence of putrefiable substances, if excluded from these same bacteria, gives rise to no putrefaction nor sepsis is shown in the cases of intra- and extra-uterine pregnancy, where a dead fœtus is carried for months or years.

The prevention of decomposition and consequent septic infection from a wound can therefore be accomplished theoretically in any one of three ways: 1, the exclusion of putrefiable materials, *i.e.*, cleanliness; 2, the exclusion of bacteria; 3, the addition of a substance in whose presence putrefactive bacteria are inert.

It is evident that the accomplishment of any one of these three ends is antiseptic, or, if you prefer on etymological grounds, aseptic surgery. There is a prevalent inclination to consider Listerism and antiseptic surgery as synonymous terms; and to regard the success in avoiding sepsis which is secured by other methods—the open air and simple water dressing, for example—as proof not only that the Listerian details are unnecessary, but also that the agency of bacteria in the induction of sepsis—an agency which the Lister method was devised to defeat—is a myth, a mere craze, a fashion. It is manifest, however, that antiseptic surgery is far more comprehensive than Listerism. Listerism aims chiefly at but one of the three possible ways for the prevention of sepsis—the exclusion of ferments; the very methods whose success has been considered proof of the fallacy of antiseptic surgery demonstrate practically what is self-evident theoretically, that putrefaction and putrid infection from a wound can be prevented by the removal of putrefiable materials, just as certainly as by the exclusion of organisms. The aseptic success of Savory and Lawson Tait—rivalling that of Volkmann, Esmarch, and Lister, was secured by the most scrupulous care in avoiding the retention or accumulation of any discharge in the wound. The result is asepsis, the means aseptic.

That this method of preventing sepsis affords the same certainty of success and possesses the same range of applicability as the Listerian, I would not maintain; indeed, my own limited experience, including some observation of surgery in St. Bartholomew's Hospital, inclines me to the contrary belief. I would merely protest against the not infrequent assertion that Savory's and Tait's success in avoiding putrid infection is an argument against the demonstrated agency of bacteria in the induction of sepsis.

There still remain a considerable number of cases, notably wounds of mucous membranes, in which anatomical relations prevent the execution of either of these aseptic methods: bacteria cannot be excluded, nor perfect cleanliness of the wound secured. In such cases asepsis can be theoretically obtained very simply by the presence of some substance in the wound which renders vital activity of bacteria impossible. There is a great variety of such agents—alcohol, carbolic acid, etc.—but for these cases all such are, from their volatility or solubility, practically useless; and it was reluctantly admitted on all sides that operation wounds involving mucous membranes could not, generally speaking, be rendered aseptic with certainty. Between 1860 and 1880 Billroth performed the amputation or extirpation of the tongue one hundred and nineteen times on one hundred different individuals; and notwithstanding the most careful attention, including frequent syringing with solutions of potassium permanganate, carbolic acid, or other antiseptic, twenty-six of these patients died, nearly all from septic infection, either directly from the wound or indirectly through the inhalation of septic products—"schluckpneumonie." With the introduction of iodoform into surgery the long-sought substance was found—comparatively insoluble and non-volatile—in whose presence the ordinary bacteria do not multiply. Under the proper use of iodoform wounds of mucous membranes are as secure from decomposition and septic infection as an amputated stump under a Lister dressing. This is admitted even by the fiercest opponents of iodoform—those who, like Kocher of Bern, having ignorantly poisoned their patients with it, would transfer to the agent the odium which evidently belongs to themselves. In 1880–81 Billroth made eighteen tongue extirpations, packing the wound with iodoform gauze which was allowed to remain undisturbed five to seven days, then sometimes renewed. Not a single septic infection occurred; recovery followed in every instance.

An operation which, though not *per se* formidable,

had, even in Billroth's skilful hands, been followed by a greater mortality than that attending ovariotomy, was deprived of its septic terrors. I would particularly recommend this to the consideration of those who ignore all experimental work, who admit as worthy of consideration only clinical results, and who regard the success of Savory and Tait as the overthrow of aseptic surgery in particular, and of bacteria in general. Here is aseptic surgery *par excellence*, though the spray, protective, mackintosh, and attendant paraphernalia are absent; here is the prevention of septic infection by measures which do not exclude bacteria from the wound, but simply restrain their development. The method of Lister, conceived and devised upon a hypothesis, before the assumptions of that hypothesis had been verified, contained, as subsequent developments demonstrated, some errors of conception and execution. The spray, for example, that sign-manual of Listerism in the professional mind, seems less essential since we have learned that bacteria are less numerous in the atmosphere than was formerly supposed; and when we consider the researches published by the German Health Bureau, it seems somewhat doubtful whether the carbolic acid spray ever killed a single healthy bacterium; the vitality of certain spores is certainly not thereby affected. Koch found that the spores of anthrax bacilli, for example, retained their power of development after immersion for seven days in a two per cent., and after twenty-four hours in a five per cent. solution of carbolic acid; yet the bacilli lost their vitality after two minutes' contact with even a one per cent. solutlon. The clinical results also support the assertion that irrigation of the wound accomplishes quite as effectually the object for which the spray was designed. Yet it must be admitted that except in those cases, such as abdominal sections, where the spray causes positive and decided injury, there is a *possibility* of benefit from its use.

For application to the wound, many substitutes have been proposed for the objectionable carbolic acid:

Fischer employs naphthalin; Schede and others report excellent results from corrosive sublimate; Langenbeck and Billroth regard iodoform as satisfactory.

To secure cleanliness—freedom of the wound from all putrefiable materials—surgeons now more generally appreciate the importance of ligating or twisting every vessel, however small, which could bleed when, with the discontinuance of the anæsthetic, the heart's impulse becomes stronger. The application of a firm, even, elastic bandage over the lips of the wound is often used, also, to accomplish the same object. For absorption of putrefiable materials Esmarch has used with great satisfaction turf enclosed in gauze bags; Schede is pleased with sand, previously heated and soaked in corrosive sublimate solution, which is poured directly into the wound. Perhaps one of the most important of antiseptic measures is the deep closure of the wound; whenever the lips of the wound are thick—as in abdominal sections and thigh amputations—the use of silver wire and lead plates for approximation of the deeper surfaces is essential to prevent the accumulation of blood in the pockets otherwise present, and the consequent danger of sepsis.

Antiseptic surgery, then, is not comprised in the spray and carbolic acid; it is not simply a question as to the relative anti-bacterial properties of this, that, and the other so-called antiseptic agents. It is an attempt to prevent the entrance into, as well as the formation within, a wound of all substances, organized and unorganized, which can interfere with cell-nutrition. It comprises, first, the exclusion or removal of all putrefiable materials—blood, pus, necrosed tissue (a point to which the Listerian school seems inclined to ascribe a subordinate place; witness Cheyne's "Antiseptic Surgery"); second, the exclusion of all ferments, bacterial or other; and, since neither of these can always be accomplished, since even under the most perfect Lister or other dressing, both putrefiable materials and bacteria may be present; third, the establishment of conditions incompatible with bacterial development. The most complete antisepsis is evidently not that which sees in bacteria the sum and substance of all surgical evil, but that which recognizes and endeavors to avoid *all* possible sources of infection. The most perfect realization of this ideal which it has been my fortune to witness is seen, not in King's College Hospital, but in Billroth's clinic. Sponges are prepared by the abstraction of fat and sand, and by at least fourteen days' immersion in five per cent. carbolic acid solution, in which they remain until used; for the operation they are put in two per cent. solution; ligatures (Billroth generally uses silk) are also kept in a similar solution. The skin at and around the location of the proposed incision is shaven, scrubbed with a flesh-brush and soap, and washed with carbolized water; hands and instruments are most scrupulously cleansed; operator and assistants wear clean linen dusters; no spray is used. Every bleeding point, however small, is caught temporarily in a clamp forceps, and at the close of the operation, ligated at the end of the severed vessel, to diminish the amount of necrotic tissue; the surface is thoroughly irrigated with three per cent. carbolic solution; a little powdered iodoform is often dusted into the wound—not, however, if immediate union be expected. If the soft parts severed be thick, the lips of the wound are approximated deeply by silver wires, and superficially by closely set silk sutures. The Esmarch bandage is removed from the limb—in amputations—as late as possible, since absorption does not occur so long as the bandage remains, but begins very actively so soon as the circulation is restored. A strip of iodoform gauze, usually also some powdered iodoform, is applied to the seam; then several layers of iodoformed or carbolized gauze; finally a very firm roller, starched organdine, or even elastic bandage is tightly applied over the lips of the wound. The first dressing remains unmolested as long as possible, the time varying, of course, with the case.

## Lecture III.

In 1879 Neisser made the assertion, based upon numerous examinations, that there is present in the purulent discharge of gonorrhœa, whether from urethra, vagina, or conjunctiva, a micrococcus not found in other pus, distinguished by its size, shape, and mode of reproduction. Neisser's previous work entitled this assertion to respectful consideration, and it was at once subjected to extensive tests. The reports have been, with one exception, unanimous in corroborating Neisser's assertion in all its details. I may mention especially Ehrlich, a most expert and experienced, yet conservative and trustworthy observer; Gaffky, a pupil and present assistant of Koch; Aufrecht, of Magdeburg; Löffler, Leistikow, Bockhart, Krause; and among the ophthalmologists, Leber, Sattler, and Hirschberg. The only dissenter, so far as I know, is Dr. Sternberg, who asserts that this micrococcus form is widely distributed, and is, in fact, the same as that which Pasteur has shown to cause fermentation of urea.

Several attempts have been made to inoculate human subjects—since animals are not susceptible to the contagion—with the isolated micrococci. Bokai, in Pesth, asserts the induction of urethral gonorrhœa in three out of six students so inoculated; but as he neglected to keep them in solitary confinement during the trial, the experiment is not so convincing as it might be. Bockhart, having cultivated the organisms on gelatine, inoculated with the fourth culture a paralytic hospital patient, and observed a typical gonorrhœa on the sixth day. Sternberg cultivated micrococci from gonorrhœal pus in flasks, and observed only negative results in each of five patients inoculated therewith. Thus far, therefore, it is not decisively established that the bacterium associated with gonorrhœa is the cause of the disease. Dr. Sternberg's present experiments, like all his previous work, evince great care, skill, and a sincere desire for truth that cannot be too much admired; yet his deductions would be far more convincing if he would substitute a solid for the liquid culture medium.

Sattler has recently found micrococci, apparently identical with those of gonorrhœa, in the conjunctival granulations, and affirms that inoculation with the organisms, isolated by cultivation, induced the disease in a human subject.

Micrococci, then, exist in the human body, locally and generally; yet excepting gonorrhœa there is no decisive evidence that a specific micrococcus is associated exclusively with any one specific morbid process in the human subject. But I would again remind you that many of these organisms are individually so minute that absolute, and hence comparative, measurements cannot be as yet accurately made; further, that micrococci morphologically identical may be physiologically distinct. Hence it cannot be asserted at present that the same species is present in septicæmia, pyæmia, etc., although the contrary is not yet established beyond doubt.

Turning to the other tribes of bacteria, however, we find more definite information; for in size, shape, mode of propagation, often of locomotion also, they present such differences that a distinction into species is often possible.

The disease variously designated anthrax, splenic fever, malignant pustule, woolsorters' disease, charbon, and by the Germans Milzbrand, is proven to be not only associated with, but also caused by, a bacillus. About this all controversy has ceased; inoculation with the bacilli, isolated by filtration, flask cultures, by cultures upon solids, by scores of observers, have always and invariably given the same result; Koch has even induced the disease by inoculation with the one hundred and fifteenth successive culture upon solids. Further experimentation is as unnecessary as further proof that a dog can be poisoned with strychnine. Anthrax is as yet the only disease proven to be due to a bacterium, by demonstrations so clear and unequivocal as to convince skepticism and silence sophistry. It is, therefore, the rock of ages on which the bacteriologists seek refuge from the waves of ridicule; the cross to which they cling amid the storms of adverse criticism; the strong castle from which they repel the impotent assaults of their enemies. The knowledge of certain facts as to the occurrence of this disease has extreme value for those of us who, having no prejudices nor views to protect, belonging to no camp nor sect in pathology, are actuated by a desire, not to demolish every one whose views do not accord with ours, but to ascertain and interpret intelligently all facts bearing upon the relations of bacteria to disease.

Anthrax is endemic in some parts of Europe, particularly of Russia, Germany, and France, and exists also in the United States. A conception of its extent may be derived from the fact that in one Russian district alone there perished in 1867–70 fifty-six thousand domestic animals—horses, cows, and sheep—and five hundred and twenty-eight human beings. In 1770 there occurred an epidemic in the West Indies, in which, within six weeks, fifteen thousand men died from eating beef infected with this parasite (Law). Sheep appear to be the natural host of the bacillus, since they are affected during the entire year, while other animals exhibit the disease only sporadically. Anthrax is both contagious and infectious; is acquired by cattle in grazing in certain localities, particularly after inundations, and in spots where animals similarly infected have grazed; may be probably acquired through the agency of flies. By man the disease is contracted through contact with infected animals, flesh, hides, wool; by eating infected meat.

These clinical facts were established long before the discovery of the bacillus; and have become intelligible and coherent only since the life-history of the parasite has been studied—for it is demonstrated that this plant produces spores, which, when placed under favorable conditions, grow into the mature form; but which, meanwhile, may remain in this embryonic state for an indefinite number of years, unaffected by extremes of temperature, by many chemical agents, even absolute alcohol. The mysterious and inexplicable sporadic appearance of the disease is at once explained. These spores may be transported in the hide, the wool, and the flesh of the animal, either of which may, therefore, cause

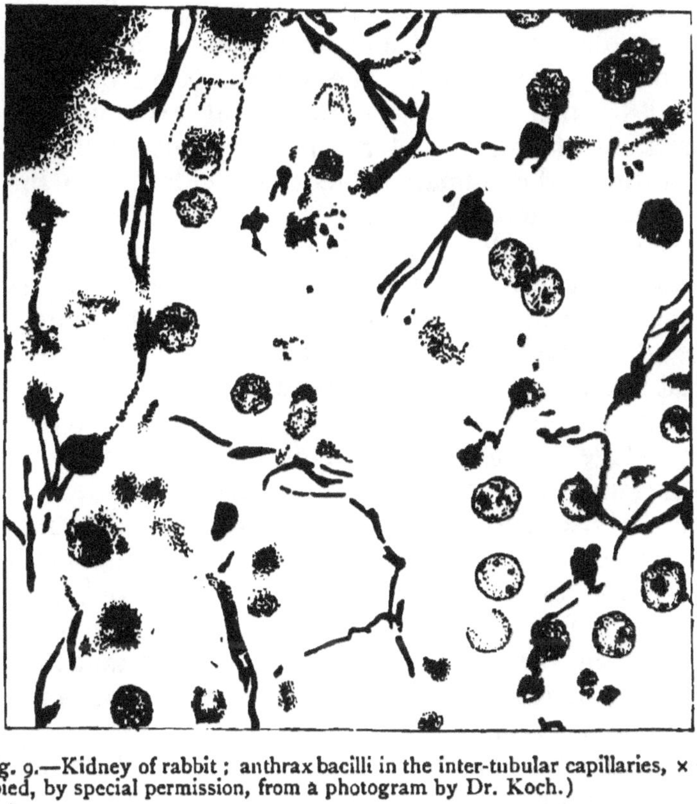

Fig. 9.—Kidney of rabbit; anthrax bacilli in the inter-tubular capillaries, × 700. (Copied, by special permission, from a photogram by Dr. Koch.)

an outbreak of the disease in a distant part of the world, and after the lapse of years. It is reported that anthrax once appeared among the workmen in an upholsterer's shop, limited to those who were engaged in repairing certain chairs, imported from a distance, which were stuffed with horse-hair. Some time ago a whole family in Scotland acquired the disease after eating soup made from beef by several hours' boiling. A local epidemic occurred in Paris among the workmen who had handled a cargo of hides from South America.

The natural habitat of the anthrax bacillus has not yet been discovered, though the plant is evidently indigenous only in certain limited districts. Koch's researches, as well as clinical observations, make it extremely probable that the bacillus is not properly a parasite of animals; that it, like many other fungi, grows upon living or dead vegetable matter, and its entrance into the cow or sheep is merely incidental to the consumption of its host as food by the animal—an accidental excursion from its usual life history—just as the presence of the trichina spiralis in the human subject is incidental to the consumption of uncooked pork.

When we reflect upon the close clinical resemblance between anthrax and certain other infectious diseases; their occurrence sporadically and epidemically; their usual limitation to certain conditions of climate and of soil; their especial prevalence during certain seasons of the year; the predisposing influence of heat and moisture; the stage of incubation; the contagiousness; the self-limitation of the disease, etc., it becomes evident to every one whose cerebral functions are normally performed that there may be, in this matter of bacteria, vastly more than the optical delusions of a microscopist, the impractical fancies of a pathologist; more than fat-crystals and fibrin threads. Yet it is understood that there may be no conclusions by analogy. Anthrax and septicæmia *may* be very similar clinically and anatomically, yet the demonstrated parasitic origin of the one does not prove the same for the other; arguments of that sort have no place in exact science. The matter must be investigated in the case of each disease independently, precisely as it has been in anthrax—a fact which is insisted upon by no one more persistently and emphatically than by Koch—to whom, by the way, we are indebted for most of what we now know about the life-history of the bacillus anthracis. And just here is another of those vital differences which distinguish Koch's work from that of Klebs and of Pasteur. The latter seem to assume the parasitic origin of the infectious diseases, and their deductions are but too often partially based upon such assumption. Koch assumes nothing, furnishes ocular demonstration of his assertions and uses all his influence, by precept and example, to raise this subject of bacterial investigation from the mire of uncertainty, doubt, skepticism, and contempt to the firm basis of exact science. For our patience has been so sorely tried, our confidence so often abused, that we have acquired a certain indifference to bacterial discoveries; we often fail to discriminate according to the evidence furnished, and regard all alike as essentially uncertain and obscure.

In this failure to discriminate between evidence and evidence, between assertions and assertions; in this failure to distinguish between a *deduction* and a *demonstration*, is to be found, in part at least, the explanation of the remarkable attitude, or rather variety of attitudes, maintained by the medical public of our land, and of our land only, on the question to be next discussed—tuberculosis. All pathologists worthy of the name, and I believe all others also, are agreed that the miliary tuberculosis of man is anatomically identical with the disease caused by the same name in rabbits, guinea-pigs, dogs, and cats; and that pulmonary consumption results from the aggregation and degeneration of miliary tubercles.

From the earliest times there seems to have been a suspicion among medical men that tuberculosis is a communicable disease; now and then an instance was observed in which a previously healthy individual, of non-consumptive stock, became tuberculous after assuming an intimate relation—as of husband or wife—to a consumptive individual; and domestic animals, even those not particularly susceptible to the disease, such as dogs, became in some instances consumptive after close attendance upon a human subject previously so afflicted. Yet the evidence of such cases, however suggestive, was not decisive; so difficult is the proof of inoculation, so insidious and gradual the inception and manifestation of the disease, so numerous and diverse the other influences to which the individual is exposed, so impracticable the restriction of personal liberty necessary for accurate observation, that the exclusion of other possible causes for the disease has not been, and probably cannot be, conclusively demonstrated in the human subject. Clinical observations have therefore never been decisive, either affirmatively or negatively. With experiments upon animals it is evidently otherwise; and this question was, early in the history of experimental pathology, submitted to experimental investigation. In 1865 Villemin demonstrated that the subcutaneous introduction of tuberculous human tissues was followed by local and general tuberculosis in rabbits and guinea-pigs. His results were in succeeding years corroborated by Klebs, Lebert, Waldenburg, Cohnheim, Fränkel, Tappeiner, Orth, Bollinger—in short by all who made the experiment; yet not every inoculation was successful: the animals most frequently subject to spontaneous tuberculosis—especially the rabbit and guinea-pig—were found also most susceptible to inoculation; those which rarely exhibit the disease spontaneously—the dog and cat, for example—often resisted attempts at artificial induction of the disease; this was to be expected, and was considered, indeed, as clinical confirmation of the anatomical evidence as to the identity of the spontaneous and the induced disease. Tuberculosis can, therefore, according to the unanimous testimony of observers, be induced by inoculation with tuberculous tissue. But it soon became doubtful whether this unquestioned fact could be interpreted as proof that there is anything specific about the tubercle; for it is evident that if all the effects produced by inoculation with tubercle can be just as certainly induced by non-tuberculous materials, no assumption of specific nature is necessary. It was demonstrated by Burdon-Sanderson, Wilson Fox, Martin, Waldenburg, Cohnheim, Fränkel, that after the introduction of mechanical or chemical irritants—a piece of wood or paper, a linen thread, a cork, glass, pepper, cantharides—in short, after the induction of irritation and inflammation in the subcutaneous tissue or peritoneum, an eruption of miliary tubercles, indistinguishable histologically from those following inoculation with tuberculous matter, often occurred. It is a little strange, by the way, that Dr. H. F. Formad, in a recent paper, called "The Bacillus Tuberculosis," in which he relates the repetition of these experiments by himself and by one of his own pupils, makes no allusion, direct or indirect, to this work to which I have just referred. This is doubtless an unintentional oversight; yet in consequence of this oversight, the casual reader derives the impression that a fact demonstrated by a score of observers in the last fifteen years was discovered two years ago in Philadelphia.

And then arose the school, represented among pathologists by Buhl and Cohnheim, and among clinicians by Niemeyer, who were inclined to deny altogether the specific nature of tuberculosis, who saw in the etiology of this disease only the caseous degeneration of an inflammatory product, a conception tersely expressed in the phrase—no cheesy product, no tuberculosis. Dr. Formad, after repeating these experiments, has recently arrived at the same conclusion; but, by a repetition of the singular oversight already mentioned, he conveys the impression, by his failure to mention Niemeyer, Buhl, and the rest (though citing one of his own pupils) that this doctrine is new.

Experimental investigation, however, revealed certain facts that demolished the Cohnheim-Niemeyer theory entirely, as admitted by Cohnheim himself.

It had long been observed that wild animals, which in their native state are not known to suffer from tuberculosis, are prone to the disease when kept in confinement; and that some tame animals, when closely confined, as is usually the case in physiological laboratories, exhibit an excessive mortality from this disease. Klebs suggested that the successful induction of tuberculosis after the insertion of glass, wood, etc., might after all be simply infection from contact with animals already tuberculous, or from tuberculous materials left in laboratories by previous subjects of the disease. Cohnheim repeated his experiments on rabbits and guinea-pigs, isolated both from other animals and from the stalls in which animals had been previously confined. The subcutaneous or intraperitoneal introduction of mechanical and chemical irritants was under these circumstances followed by tuberculosis—in *not a single instance.* Fränkel, who had performed with Cohnheim the original successful experiments in the Berlin Pathological Institute, repeated them in his private dwelling, with *absolutely negative results.* Cohnheim, with the moral courage born of true scientific spirit, published this fact, and acknowledged the justice of Klebs' suggestion.

Chauveau, Aufrecht, Bollinger, and others proved that tuberculosis can be induced in rabbits and other animals by simply mixing with their food tuberculous material, and that this tuberculosis begins not in the lungs, nor in some caseous inflammatory product, but directly in the intestinal wall. Giboux placed healthy rabbits in cages in each of two rooms; into one room was passed, for several hours a day, the breath expired by phthisical patients; into the other room the same, after filtering through carbolized tow; in a few months the rabbits in the first room were dead of tuberculosis; in the second apartment there was no sign of death nor of tuberculosis. Tappeiner, and after him Bertheau, demonstrated that the inhalation of sputum from phthisical patients in minute quantity is followed by pulmonary, and then general tuberculosis, not only in rabbits, which are so susceptible to the disease, but even in dogs; and that the inhalation of other sputum did not produce this effect. On the other hand, Schottelius observed among the inflammatory products following the inhalation of irritant particles, such as malodorous cheese, certain nodules histologically identical with spontaneous tubercle; and that similar nodules sometimes followed the prolonged inhalation of non-phthisical sputum in large quantity. (Schottelius, by the way, has since acquiesced in the infectiousness of tuberculosis.) Weichselbaum found a few similar nodules in the lungs of one of three dogs treated in this way, but no tuberculosis. He found further, that the inhalation even when brief, of phthisical sputum induced general tuberculosis; but after boiling, or after treatment with corrosive sublimate, the same sputum produced no tuberculosis, and rarely if ever nodules.

These inhalation experiments—of Tappeiner, Bertheau, Schottelius, and Weichselbaum—illustrate admirably from the experimental side what had been for years acknowledged from the histological standpoint: that there is nothing characteristic in the individual nodule. The same histological structure, including the giant-cell, may be found in the nodules of syphilis or lupus, as well as of tuberculosis; the same local anatomical change may follow the inhalation of large particles of Limburger cheese, as the inspiration of atomized phthisical sputum. For decades the pathologists from Virchow down, their eyes full of caseous matter and giant-cells, wrestled with one another over the question, what is true tubercle? Their hair-splitting disputes remind us of the bitter controversies of the mediæval philosophers as to how many spirits could stand on the point of a needle. Finally it dawned upon them—they were confounding anatomy with etiology; they were regarding as characteristic of one morbid process a histological structure common to several; they were ascribing to a single cause the common effect of many; *they were confounding tubercle with tuberculosis.* As Cohnheim said years ago: "Struggle against this as we may, there is no help for it—the anatomical definition suffices no longer for the tubercle and tuberculosis." Even Schottelius, the last of the German pathologists to deny the infectiousness of tuberculosis, has finally recorded his conviction that tuberculosis is certainly infectious, though not all individual tubercles belong to tuberculosis.

And what shall be considered a tuberculous tubercle?

Wherein may we distinguish a "true" tubercle from a nodule exhibiting an identical structure? The question is answered instantly when we consider what constitutes pyæmic pus. Pus is a definite anatomical entity, varying, like tubercle, within certain limits. The pus from a pyæmic joint, may be indistinguishable by the microscope from the pus of a simple abscess; yet there is none the less a vital, or rather a fatal difference. Fresh pus from a simple non-specific abscess does not cause pyæmia, as Virchow long ago proved; a minute quantity of pyæmic pus is fatal through pyæmia, as the death of many a physician has testified. The tubercle from Limburger cheese does not cause tuberculosis, as Schottelius himself admitted; the tubercle from tuberculosis *never fails* to do so, as all observers testify. Pyæmic pus, however similar histologically to that from croton-oil or turpentine, is unerringly distinguished by its infectiousness; tubercles from tuberculosis, though anatomically identical with those from mechanical irritation or from syphilis, contain a something sure to propagate tuberculosis in the proper soil. The non-specific pus of a simple wound or abscess may acquire pyæmic properties, without the intentional or even conscious introduction of pyæmic matter; the non-specific cheesy products of a simple inflammation may acquire infectiousness without the *intentional* introduction of a specific agent. This acquisition of pyæmic infectiousness never occurs, as has long been known, without bacteria; and the day has come when we can say that the infectious tubercle—of tuberculosis—is also characterized by a bacterium.

Perhaps the clearest proof—because a demonstration—of the infectiousness of tuberculosis is furnished by comparative observations upon the eye made by Baumgarten, Cohnheim, Salomonsen, Deutschmann, and others. When the piece used for infection is fresh, Cohnheim says "the irritation at the commencement usually soon subsides, the particle becomes gradually smaller, and can indeed entirely disappear, and for some time the eye then appears entirely clear and intact, until there suddenly appears in the iris a greater or less number of very minute gray tubercles which, precisely like the human tubercles, grow to a certain size, then become caseous, etc. In rabbits Salomonsen and I observed the eruption of the tubercles usually about the twenty-first day after the inoculation, in guinea-pigs a week earlier as a rule." "Yet these observations have first acquired their full significance from the fact that the tuberculosis is generated by the inoculation of tuberculous matter *only* and of *nothing else*." These are the words of a man who, some years ago, was inclined to the belief, from the fact that the disease may follow subcutaneous or intraperitoneal mechanical irritation, that tuberculosis was not an etiologically specific disease. In his "Allgemeine Pathologie," in discussing the same subject, Cohnheim says, "After a few days the cornea is quite clear, the iris thoroughly clean and in perfectly normal condition; in the aqueous humor there is also no exudate to be seen, so that one can see the piece introduced sharply and clearly defined against the lens capsule; and thus it remains unchanged for weeks, except perhaps that the particle becomes somewhat smaller. All at once, in our cases between the twentieth and thirtieth day, the scene changes; there arises in the iris tissue a considerable number of small transparent grayish tubercles." "Yet the most interesting feature is that in numerous instances, though not always, a more or less extensive tuberculosis of lungs, lymph-glands, spleen, and other organs, occurs. From these experiments it cannot be doubted, first, that the tuberculosis by inoculation can develop *without the medium of a coagulated exudate*, and, second, that it has a stage of incubation." He says further, "Where experiments so positive and so easy of repetition are adduced, it would seem impossible to discuss any longer the question of infectiousness."

Until, therefore, we can otherwise explain the fact that general tuberculosis can be induced by mixing small quantities of tuberculous matter (but not by mixing any other tissue) with food; by simple inhalation of phthisical sputum, but by no other sputum, nor even by this after boiling or treatment with corrosive sublimate or filtration through carbolized tow; that general tuberculosis without caseous exudate follows the introduction of a minute tuberculous particle, but of nothing else, into the eye; until it shall be possible to offer another explanation, we must admit that there is a *something* peculiar to tuberculosis and not common to all tubercles. To assert with Niemeyer that the disease originates *de novo* in a cheesy mass, is to assume that because there is no *intentional* or *conscious* introduction of an infectious agent therefore none occurs. Surely no surgeon ever intentionally or consciously introduced pyæmic matter into a wound; yet infectious pyæmia was formerly the scourge of hospitals. Septicæmia, erysipelas, diphtheria, and pyæmia are none the less infectious because there is—especially in the so-called spontaneous cases of each—no discoverable possibility of contact with previous subjects of the same disease. *Infection*, in other words, does not necessarily imply *contagion.* No man becomes syphilitic unless there be incorporated into his body material from an individual previously syphilitic; no man acquires scabies without contact with a sufferer from itch; but pyæmia, erysipelas, diphtheria, anthrax, and tuberculosis are acquired not only by transfer from subjects of the respective diseases, but also without such transfer. The purulent secretions of a wound are doubtless favorable soil for the retention and propagation of pyæmic or erysipelatous virus; the catarrhal products in the throat for the origin of diphtheritic infection; the cheesy products of a bronchitis or of mechanical irritation for the location and propagation of the tuberculous infective agent; *but none of these are necessary.* Pyæmia, erysipelas, diphtheria, anthrax, tuberculosis occur not only by demonstrable contagion, not only after a simple wound without demonstrable contagion, but also *without either demonstrable contagion or even a simple wound—i.e.*, spontaneously.

Yet in the face of this perfect analogy with other infectious diseases, in the face of the experimental proof as above related, there are doubtless many practising physicians who cannot believe that tuberculosis is communicable. Why? First, because clinical proof to that effect is unsatisfactory; a surgeon pricks his finger in dressing a pyæmic patient, and in twenty-four hours has a chill and local symptoms pointing unmistakably to the source of infection; a physician inspires the breath of a struggling diphtheritic patient, and in three or four days gives evidence that the disease was communicated. Had tuberculosis ever been observed to occur in animals so soon or so violently, were the introduction of tuberculous as well as of pyæmic and erysipelatous virus accompanied by chill, fever, and severe acute local inflammation (it does seem to be so accompanied in general acute miliary tuberculosis), there might be reason in the objection that no absolute clinical proof has been furnished. But even when the freshest of tuberculous material is introduced into the most favorable soil, the eye of a susceptible rabbit, two to four weeks elapse before the first local manifestation of infection, and further weeks or months before the evidences of general tuberculosis are apparent; there is, indeed, nothing in the animal's history to indicate infection, the proof of which consists merely in the conscious act of inoculation. An observer who was not aware of this act, might honestly believe that the infection which manifests itself weeks or months later is spontaneous—even autochthonous; and some physicians, because they see no transfer of tuberculous material (though opportunities enough for such transfer are certainly given), because they see *no striking symptoms to mark the hour or the day of infection*, insist that no infection has occurred.

A man may be shot in the presence of witnesses; but if we find a body with a bullet in the heart, we are none the less certain that this body, alive or dead, has been shot, though no revolver nor human agent may be discoverable. A man may be killed by the lightning which dazzles all eyes; but he is none the less killed if the electricity be the invisible current of a powerful battery.

If a man exhibits secondary syphilis, nothing can shake our conviction that he has, whether consciously or not, come into contact, mediate or immediate, with a syphilitic person, although the sufferer himself may be honestly unable to point out the moment or the mode of possible infection. In many cases of pyæmia and diphtheria the course of the infection is as plain as the track of the lightning; the effect as pronounced and almost as sudden as that of the murderer's bullet. In other cases, however, of pyæmia, diphtheria, as well as of anthrax, syphilis, and tuberculosis, there are absolutely no phenomena observable in the individual which attract attention to a given moment as the time at which an infection, subsequently manifested, may have occurred. A piece of tuberculous matter introduced into a rabbit's eye may entirely disappear; and for weeks the animal presents absolutely no signs, local or general, of tuberculous infection.

The opportunities for the usual mode of infection by syphilis are only occasional; and the attendant circumstances are such as to impress such occasions upon the mind and conscience; when, therefore, the first evidence of infection appears, perhaps weeks subsequently, upon that part of the anatomy peculiarly exposed upon such an occasion, it is but natural that the mind should associate the two phenomena as cause and effect. Were syphilis communicated not in the way at present in vogue, but by inhalation; were the initial evidence of infection not upon the integuments and therefore visible, but in the lungs and hence inaccessible to the eye, there might be the same clinical grounds for doubting the infectiousness of syphilis as of tuberculosis. The occurrence of infection in tuberculosis is usually as unobserved clinically as in the exceptional cases of syphilis, in which a primary lesion was neither suspected nor discoverable.

Another argument often heard against the infectiousness of tuberculosis, recently uttered and printed by a Philadelphia surgeon, is the fact that we do not all die of this disease. Yet the same argument is valid against the infectiousness of cholera, yellow fever, diphtheria, scarlatina, etc. Indeed, since the death-roll of tuberculosis is greater, year after year, than that of any one or perhaps all of these diseases combined, the argument, if it had any sense at all, would tend to prove the excessive infectiousness of tuberculosis. Such an argument ignores the unquestioned and familiar fact that we are not all equally susceptible to any one of the infectious diseases; even the most malignant cholera or yellow fever attacks only a portion—usually a decided minority—of the community. Explain it as we may, there is a something which we may call predisposition, by virtue of which only certain individuals yield to infection by cholera or by tuberculosis; and the fact is, that the number susceptible to tuberculosis seems smaller than to any one of several other infections. Comparatively few of us attain maturity without having had measles, scarlet fever, and whooping-cough at least; yet six-sevenths of us complete our pilgrimage without exhibiting evidences of tuberculosis. That this is not mere accident is shown by experiment: guinea-pigs and rabbits rarely, dogs and cats usually fail to respond with general tuberculosis to inoculation with tuberculous material. Even the deadly anthrax usually fails to destroy carnivorous animals, although the most virulent material be introduced; and it was long ago pointed out by Chauveau, and often confirmed, that although sheep are very susceptible to this disease, yet some sheep resist all experimental attempts at inoculation, even when large quantities of fresh anthrax material are injected into the animal. Dogs enjoy in general immunity against infection by anthrax; yet young dogs are often successfully inoculated. Infection implies, therefore, not simply a virus capable of propagation in an animal, but also an animal capable of permitting such propagation. All variations of this relative adaptability may be exhibited between animals of the same species and a given virus. To affirm, then, that a disease—anthrax or tuberculosis, for example—is infectious is to assert that it can be communicated by the diseased to *a* healthy animal, not to *all* healthy animals, even of the same species. Herein lies evidently our security against tuberculosis, as well as against many other infectious diseases. The general principle—the survival of the fittest—seems to have been for generations at work in eradicating this disease from the human family, by removing those members of it susceptible to tuberculosis; the great majority of us now living are as safe from tuberculosis as most dogs are from anthrax.

Another, perhaps the most profound, argument against the infectiousness of tuberculosis should be considered here, namely, that the fact must cause us to relapse into barbarism. "Some of the most noble and tender traits of humanity threaten to be undermined. The consumptive who has been heretofore lavishly loved and cared for," etc., "is to be isolated and shunned as a leper, if such doctrines prevail" (*Philadelphia Medical News*, January 27th, p. 94). Therefore tuberculosis is not infectious. Incredible as it may appear, the author of these lines is not a clergyman nor a poet, but a distinguished surgeon who does not shun infectious pyæmia, septicæmia, and erysipelas; and who, we may assume, does not love his child less lavishly, nor care for his patient less faithfully because that child or patient may suffer from infectious diphtheria or scarlatina.

Were the susceptibility to tuberculosis as general as to diphtheria, scarlatina, and measles, there might be grounds, not for shunning the consumptive "as a leper," but for the observance of proper precautions for the protection of the healthy many, even at the inconvenience of the diseased few. But since the experience of generations has shown that only about one-seventh of us acquire tuberculosis even with unrestrained intercourse with consumptives, it may be questionable whether any other protection than a knowledge of its infectiousness for some individuals be necessary; we do not invoke the law to brand syphilitic individuals, though to this infection not one-seventh but, probably, all of us are susceptible. But, however that may be settled, let us not confound a fact with a possible deduction which may be unpleasant.

These more or less prevalent arguments against the infectiousness of tuberculosis have been considered not because they have any bearing upon the question, but because there are those who will not or do not take into consideration the demonstrations attained by accurate experimental methods, and whose opinions rest upon distorted deductions from necessarily inaccurate clinical observations. Yet while those who are pleased to regard pathology as something extrinsic to practical medicine are still discussing the clinical proofs of the infectiousness of tuberculosis, it is quite otherwise with pathologists and clinicians whose opinions are founded upon knowledge without prejudice. One after another the German and French pathologists (who are not infrequently clinical teachers as well), honest in their previous conviction that the communicability of tuberculosis was not proven, honestly recorded their convictions as succeeding proofs were furnished, that the case was reversed; so that three years ago Cohnheim said, "To-day there scarcely exists a pathologist who would deny that tuberculosis is a communicable disease."[1]

Cohnheim himself, extending and repeating more carefully his observations, saw and acknowledged the error of his former deduction. True, a would-be pathologist has occasionally reminded us that he was not yet convinced; yet even Schottelius, the last of them, has finally yielded the point. There have been in all ages, and on all questions, similar psychological curiosities; twenty-five years ago it was maintained on the floor of the French Academy of Sciences that intestinal worms originate *de novo* in a peculiar influence pervading the system—the vermicular diathesis. There is a gentleman in this State who recently reminded us that bacteria, so-called,

---

[1] In the Medical News, January 27, 1883, p. 94, Dr. Wm. Hunt leads us to infer that "most recent pathologists" agree in regarding tuberculosis as the result of a simple inflammation. Will he kindly name *one* pathologist who now holds this opinion, and mention the pertinent publication?

are in his opinion fibrin threads and the like; and there is said to be a man in Virginia who still insists that the earth is flat.

You may have noticed that in this discussion the name of Koch has not been mentioned—a fact to which I call attention, because a popular impression, not entirely confined to the laity, saddles upon Koch the paternity not only of the bacillus, but also of the infectiousness of tuberculosis. Dr. Formad, for example, says (p. 3): "An analysis of Koch's experiments shows that he has not proved the parasitic nature of tuberculosis, so *that the infectiousness of tubercular disease is still sub judice.*" It is apparent from the facts which I have endeavored to summarize that the communicability of tuberculosis was established years before the well-known publication of Koch's discovery. Dr. Formad says (p. 10): "The supreme question before the medical world is now, whether the disease under consideration is really infectious." This statement may represent faithfully that portion of the world bounded by the city limits of Philadelphia; the supreme question before that portion of the medical world including Virchow, Cohnheim, Billroth, Bamberger, Weigert, Villemin, and the other German, French, and Austrian pathologists and clinical teachers is, *not* whether tuberculosis is infectious, but whether the bacillus of Koch is the infective agent. For them the two questions are quite independent—the former established, the latter awaiting confirmation.

The numerous examinations of tuberculous tissues revealed occasionally bacteria, which the discoverers were but too willing to consider the cause of the disease; Klebs, Schüller, and Aufrecht severally announced but failed to demonstrate that the infective agent had been found and that it was a bacterium. The lack of evidence in support of their statements, as well as the reserve with which such assertions in general were received, combined to reduce to a minimum the attention bestowed upon them. Such was the state of affairs when Koch read before the Physiological Society of Berlin a paper whose contents were in forty-eight hours telegraphed over the world. Koch's statements are so familiar to all, that detailed repetition would be superfluous; they may be summarized in the assertion that the active agent in the induction and propagation of tuberculosis is a distinct species of bacterium, a bacillus; that tuberculosis does not occur without the presence of this organism; that conversely all those anatomical changes and only those should be called tuberculosis whose point of departure from the normal condition is the presence and vital activity of this bacillus; hence, general and local miliary tuberculosis, cheesy pneumonia and bronchitis certainly, fungus-joint granulations, scrofulous inflammation of lymph-glands probably, and the pearl disease of cattle, are etiologically identical. The point of chief interest is of course the assertion that tuberculosis and cheesy pneumonia, pulmonary consumption, are caused by the bacillus. The evidence in its favor is first the experimental work of Koch himself, and then the unanimous confirmation of those of his statements which have been already tested.

His experience with and knowledge of bacteria found in the animal body is by general consent admitted to be excelled by that of no other observer; his caution and conservatism and the accuracy of his methods are such that, although he has for eight years been constantly working and frequently writing on this subject, he has never as yet been detected in a single error of observation; his facilities and opportunities in the Imperial German Health Bureau are unexcelled. The confidence and good will of government and people alike—for Koch's is an official position, you know—would be destroyed by any ill-executed observations, or by any injudicious and untenable assertions in this, the most important and widely circulated of all his works. That Koch appreciated the situation is shown by his course in the matter: having discovered the bacilli in tuberculous tissue, he did not send an announcement to the Academy of Sciences nor blazon it through the medical press; he kept it to himself, satisfied himself that this was a constant, not an occasional or accidental association; that the same bacteria were present in the spontaneous tuberculosis of animals—the hog, chicken, ape, guinea-pig, and rabbit; then he devised, by experimentation, a proper medium, *solid* of course, for cultivating the organisms outside of the animal body under constant microscopic supervision, comparing them with fresh bacilli from tuberculous tissues; satisfied himself again by personal experiment of the inoculability of tuberculosis; found that while vaccination of the rabbit or guinea-pig with fresh tuberculous matter induced the disease, inoculation with such material after lying in alcohol for a month or dry for two months, was impotent to cause the disease, and *contained no living bacilli;* found that the bacteria were often, *not always*, present in the sputum of tuberculous patients, but never, so far as examined, in that of others. Having thus made a preliminary investigation, Koch proceeded to the experimentum crucis with bacilli which had grown from the tuberculous tissue under his eyes; which were therefore proved to be the progeny of the original ones, not by the theory of probabilities, not simply by their identical size, shape, and chemical reaction, but by the fact that he had seen them proceed from the first as continuities of structure; which were seen under the microscope to be quite free from any foreign solid matter, bacterial or other; which were proven to be equally free from any foreign matter in solution because growing in successive cultures upon solid soil; which had been carried from the first to the eighth generation; which had been thus isolated from the original animal tissues three, four, five, even six months. With these isolated descendants of the bacilli found in tuberculous tissues, Koch inoculated numerous animals—using over two hundred altogether—not only the susceptible rabbits and guinea-pigs, but also cats, a dog, white rats which had resisted inoculation by injection and by feeding with tuberculous materials, and field mice. Inoculation was made in the skin, the abdomen, the eye. In every case tuberculosis and tubercle bacilli were found in the infected animal.

Having spent *two years* in the completion of this work, amid all the facilities of the imperial laboratory; having meanwhile pérmitted himself no public intimation of the same, Koch quietly announced his results at a regular meeting of a medical society, with as little ostentation as if he had merely appropriated a chapter from Ziemssen. One whose knowledge of bacteria and of disease is not such as to permit a technical appreciation of Koch's work, cannot help seeing in the unobtrusive, systematic, and undeviating work of two years, and in the modest announcement of the result, that Koch's work is not to be classed with that of Klebs, or Letzerich, or even Pasteur. I would call your attention to the fact that Koch's assertion embodies not a theory, but simply an ocular demonstration. If a man is seen to plunge a knife into the heart of another the killing is a fact, not a theory; if Koch saw tuberculosis invariably follow the introduction of isolated bacilli, the relation of cause and effect is a *fact, not a theory*. There is only one possible escape (I use this word intentionally out of regard for the prejudices of many friends) from the conclusion that the bacillus causes tuberculosis; and this forlorn hope is the possibility that Koch did not see what he says he saw—that he made some vital error of observation. This is of course possible, though if true it will be the first error that the most searching scrutiny could ever detect in his observations; that it is improbable is evident.

If we accept Koch's observations as accurate, there is only one conclusion—that these bacilli cause tuberculosis. For here the conclusion and the observation are identical; this is not a deduction, but a demonstration.

And how shall it be decided that this work is or is not free from errors of observation? Certainly not by saying that it cannot be so; not by exhuming Niemeyer's buried argument that tuberculosis is not infectious; but simply and solely through the repetition, by competent observers, of the same work. Until such repetition shall detect serious errors of observation, Koch's work stands unchallenged—more accurate and complete investigation can scarcely be conceived. On the other hand, until such repetition shall confirm Koch's observations, we may justly decline to accept them unreservedly, on the ground that he *may* have made his first error in this his greatest effort.

While, however, Koch's main assertion, that the bacilli cause tuberculosis, can be competently criticised only by the few men who like himself have the time, facilities, and skill necessary to conduct such tedious and delicate experimental observations, yet some of the preliminary assertions fall within the range of a larger circle of critics, and have been already subjected to extensive investigation. The results are as yet unanimous in confirming the original assertions of Koch that the bacilli are to be found in the sputum from most though not all cases of pulmonary tuberculosis, and, what is quite as significant, have never been found in any other sputum.

Ehrlich, Balmer and Fräntzel, Guttmann, d'Espine, Lichtheim, Fränkel, Ziehl, Heron, Gibbes, Green, West, Yeo, Whipham, Councilman, have already recorded their unanimous experience that while the bacilli are found in the sputum in at least a large majority of cases of pulmonary consumption and tuberculosis, they are not found in any other disease. Balmer and Fräntzel found them always in their one hundred and twenty cases, but never in bronchitis. Ziehl recognized them in nearly all of seventy-three cases, but never in thirty-four other cases, including acute and chronic bronchitis, acute fibrous pneumonia, gangrene of the lungs—indeed all pulmonary diseases that he had opportunity to examine. It should be remembered that Koch failed to find them in the sputum from a certain number of cases. In the tubercles of tuberculosis and in the cheesy matter of consumptive lungs the bacilli are usually present—not always, as Koch himself discovered. Whether their absence from certain tubercles is to be explained, as Koch suggests, by the death of the organisms and their consequent failure to absorb aniline colors, or whether some of these tubercles arise from other causes than the presence of these, may be perhaps an open question. Gibbes' experience—that the bacilli are present in only one reticular nodule out of ten, but in nearly all non-reticular tubercles—might, perhaps, support another explanation. Certainly the absence of the organisms from tuberculous tissue is the exception. On the other hand, the bacillus is never found in the body except in tuberculosis; the only suggestion to the contrary is the recent assertion of Koranyi that he found similar organisms in a case which he believed to be pulmonary syphilis, and not consumption.

Such, then, is the state of the case to-day: Koch's assertion of the association of the bacillus with tuberculosis—its presence in every case of the disease, its absence in all other morbid conditions—confirmed by all who have investigated; his assertion of the causal relation of the parasite to the process—based upon a demonstration unexcelled in the history of experimental science for accuracy, clearness, and completeness—as yet unchallenged.

The subject might be properly left here; but I deem it advisable to consider briefly two recent publications, not because they demand consideration by one familiar with the facts, but because they may have influenced some who derive their information chiefly from American literature.

A few months ago there was heard a scream of exultation from a Western journal, soon echoed on many sides. The attention of press and public alike was attracted to the jubilant cry that Koch, bacillus, and bacteria were to be annihilated; that the "bacillary craze" of German pathologists; the absurd fancy that a small organism could harm a large one; the comical idea that an experienced mycologist should know more about bacteria than a practising physician; the barbarous doctrine that our loved ones could be subject to infectious diseases; all these and similar absurdities which pseudoscientists had vainly attempted to foist upon our superior intelligence would be forever buried. The American eagle, that implacable devourer of microscopic poultry, would consent to leave for a brief time its favorite swamp at the "delta of the Mississippi," and by a single act of deglutition would teach our terrified friends, "the micro-pathologists," to "take their eyes from their mounted specimens," and engage in less disreputable pursuits. So ran the widely advertised programme. After weeks of joyous anticipation the appointed day arrived; a distinguished microscopist, whose skill in mycology had been amply indicated by his failure to detect the bacilli always present in leprous tissue, appeared in the arena armed with the startling discovery that if caustic potash solution be added to fattily degenerated tissue, crystals of fatty acid appear! The announcement was greeted by the audience of assembled experts with rounds of applause—"Sic transit bacteria," etc. Again has free America repelled the assaults of effete Europe.

I had intended to offer some remarks upon this matter, suggested by the evident fact that Koch's bacillus and Schmidt's crystal were different objects; but criticism is no longer necessary. Dr. Whittaker has stated the case very clearly; Dr. Hunt has shown that the crystal polarizes light, while the bacillus does not; and I have received from Dr. Schmidt a letter which disarms criticism.

After reading his article, I sent him a slide of sputum containing the bacilli; in his reply he says, "From what I understand now the minute crystalline rods which I discovered are not identical with Koch's bacilli;" and later, "the failure with which I met in my attempts of staining the bacillus tuberculosis, appears to have been due to the worthless aniline oil which I have used." I interpret these sentences as a candid admission that the crystal and the bacillus are not identical, and shall therefore refrain from further remark. Such admission, by proving sincerity of purpose, transfers to his friends of the antibacterial "camp" the obloquy and chagrin consequent upon the blare of trumpets with which this publication was heralded; the mouse may be *per se* a highly respectable and by no means ridiculous animal, though its advent as the result of herculean efforts at parturition is said to be very absurd. This entire matter can hardly fail to teach far more effectually than lectures, that trustworthy investigations on this subject demand not only skill and experience in pathology, which Dr. Schmidt undoubtedly possesses, but also acquaintance with the special methods involved.

A paper called "The Bacillus Tuberculosis," by Dr. H. F. Formad, of Philadelphia (*Philadelphia Medical Times*, November 18, 1882, reprint), opens with the announcement that the author "will bring forward some points from researches of my own, which will check the acceptance of the doctrine of the parasitic origin of tuberculosis;" "my anatomical researches will also surely throw grave doubts upon the correctness of Koch's views on the etiology of tuberculosis" (p. 2). The author fails to discriminate between the bacillus and the infectiousness of tuberculosis, which is in this article, however, a matter of little consequence, except as an index to the general accuracy of the publication.

The original researches which are to destroy the "parasitic theory" consist, curiously enough, in the time-honored demonstration that tuberculosis often occurs in certain animals (notably the rabbit and guinea-pig) after simple wounds, the irritation caused by glass, etc.; especially if the animals be carefully confined in a pathological laboratory where many others have died of this disease. As Dr. Formad has seen "more than one hundred rabbits, out of five or six hundred operated upon," die of tuberculosis, we may infer that in his laboratory there was no lack of tuberculous material for infection.

There is, however, one original feature in this work as reported by Dr. Formad. Actuated doubtless by a commendable high-tariff spirit of protection for American industry, while quoting copiously his own students, he resolutely ignores the work of Burdon-Sanderson, Cohnheim, and a dozen others who have, during the last fifteen years, performed the original experiments of which his own are repetitions; and neglects to state that Cohnheim and Frānkel found that while these experiments succeeded admirably in the Berlin laboratory where many animals had long been confined, no tuberculosis occurred in a subsequent repetition in a private dwelling. On the same principle, perhaps, he neglects to state that for such reasons as these, such experiments as his own were years ago abandoned to amateurs, while the battle for infectiousness was fought and won in the eye, the lung, and the intestine, as above stated. Perhaps Dr. Formad will kindly explain how he came to deny the infectiousness of tuberculosis merely on the strength of these long since abandoned experiments, without a solitary experiment, or even reference to an experiment, on the eye, etc.

Because in his experiments no tubercular matter was "intentionally or knowingly" introduced, he maintains that nothing could have entered; that the disease is therefore not specific nor infectious. Surgeons, then, intentionally and consciously inoculate their patients with pyæmic, diphtheritic, and erysipelatous material. It will not help Dr. Formad to deny, as a New York microscopist in the same dilemma has curiously done (MEDICAL RECORD, March 3, p. 247), that pyæmia is infectious. For in the *National Board of Health Bulletin*, Sup. No. 17, Formad asserts and attempts to prove the infectiousness of diphtheria, and says (p. 18): "A case may begin as one of sthenic pseudo-membranous croup, and end as one of adynamic diphtheria with blood-poisoning; and in cases of this character, not infrequently, *no exposure to contagion is discoverable.*" [1] Perhaps he will explain why the absence of intentional or conscious inoculation, even of discoverable exposure to contagion, is perfectly compatible with the infectiousness of diphtheria, and yet proves the non-infectiousness of tuberculosis. Formad says (p. 2): "I can positively prove that true tubercu-

[1] Italics mine.

losis may be produced without the bacillus in question." The only proof adduced for this important statement is the experiment with glass, etc., in which the disease occurs without any "conscious or intentional" introduction of the bacillus, and the *assumption* that the organisms were therefore absent; if, however, the parasites be nevertheless present in such cases, this assertion is evidently unwarranted. We are not informed on this point in the paper, although we may infer their presence from the following statement (p. 11): "Koch has discovered that tubercle-tissue is *always* infested by bacilli, and this is correct."[1] To secure definite information, I addressed to Dr. Formad some months ago, three several letters enclosing stamps, requesting him to state for incorporation in these lectures, whether he had examined these cases of tuberculosis following wounds, mechanical irritation, etc., to ascertain the presence or absence of the bacilli, and if so, with what result. To these letters I have received no reply.

As to the association of the bacilli with tuberculosis, Formad's limited observations seem to agree with Koch's statements.[2]

The one novelty—which I am charitably disposed to think explains the existence as well as the peculiar character of this paper—is a theory that the susceptibility to tuberculosis is inversely proportional to the width of lymph-spaces; whence (by a process of reasoning peculiar to the author) he makes the deduction that no etiological influence other than inflammation and narrow lymph-spaces is necessary to induce tuberculosis. It is useless to remind Dr. Formad of the induction of tuberculosis in the eye, lung, and intestine, since he ignores pathological work which does not emanate from himself or his pupils. But since by a singular coincidence the rabbit and guinea-pig—the animals exhibiting typical narrow lymph-spaces

---

[1] Koch, by the way, does not make this statement, which is, moreover, *not* correct, since Koch, Gibbes, Ziehl, Guttmann and others failed to detect the bacilli in a certain number of tubercles.

[2] Paradox: "Tubercle tissue is always infested by bacilli," yet "true tuberculosis may be produced without" them.

—are peculiarly susceptible, the dog and cat insusceptible to anthrax as well as to tuberculosis, I would suggest the possibility that the etiology of anthrax also may be found, not in a bacillus as the Europeans suppose, but in narrow lymph-spaces.[1]

Dr. Formad promulgates the dogma (p. 3) "Scrofulous beings" (*i.e.*, those with narrow lymph-spaces) "can have no other than a tuberculous inflammation, although it may remain local and harmless." Are scrofulous beings, then, assured against syphilitic, erysipelatous, diphtheritic inflammations, or are these merely varieties of the tuberculous?

As yet the presence of the bacillus in sputum has possessed a confirmative rather than a diagnostic value, for in the cases in which it has been detected the diagnosis has been usually already assured by the physical signs. Whether or not the bacillus may be present in cases called chronic bronchitis, etc., where the symptoms and the family history beget a suspicion not yet supported by physical exploration, must be decided in the future. In this connection it may be proper to mention an instance which has fallen under my own observation.[2]

A young gentleman of my acquaintance, in whose family history there is no record of consumption, but who had for months suffered from a persistent and annoying cough, requested me one day to examine a microscopic slide which he had prepared. The diagnosis was easy, tubercle bacilli in sputum. He then informed me that the sputum was his own. Physical exploration by one of our most experienced physicians revealed subsequently a circumscribed area of consolidation in the right lung.

Fränkel has always found the bacilli in laryngeal ulcers of tuberculous patients, but never in those of syphilitic

---

[1] Dr. Formad will enlighten us, in subsequent "Pathological Studies," as to what he will permit us to call "true" tubercle, and announces that a student of his is incubating a cognate topic. Possibly we may yet learn what constitutes "true" pus; and how many spirits can stand on a needle-point.

[2] I have recently learned from my friend Professor W. H. Welch, of New York, that two essentially similar cases are known to him.

or other individuals. Barrow found them in the urine from tuberculous kidneys in one case.

Not only the clinical, but also the anatomical investigation already reported, confirm Koch's statement that tuberculous tissue, whether occurring in miliary nodules or as cheesy masses, whether in lung, or liver, or spleen, peritoneum, or meninges, contain tubercle bacilli, and that no other tissue harbors them. Some, it is true, find a larger proportion than did Koch, of individual miliary tubercles in which no bacilli can be detected; this is particularly true of Ziehl's examinations. Gibbes found the bacilli in reticular tubercle in only one nodule out of ten, in the non-reticular they were usually present. Koch was inclined to the belief that his failure to detect them was due to the fact that the organisms had lost their vitality, and hence their power of absorbing aniline dyes; and demonstrated instances in which a very imperfect staining of individual bacilli was visible. This explanation is certainly plausible, yet it is possible that tuberculosis, like individual tubercles, may be produced by any one of several causes. The clinical picture exhibits many variations; the histological structure is not peculiar to tubercle. We have learned to distinguish trichinosis from typhoid fever; charbon symptomatique from charbon; actinomycosis from pyæmia and pulmonary consumption; indeed, Pflug observed in the lungs of a cow a miliary tuberculosis, and upon microscopic examination was surprised to find that the individual tubercles contained, not Koch's bacillus but the *actinomyces bovis;* this as yet solitary observation strengthens the suspicion that among the numerous agents whose presence excites the inflammation which results in tubercle formation, there may be other parasites than the bacillus of Koch; that there may be several diseases etiologically distinct, but anatomically so similar as to be included under the common name tuberculosis; the one characterized by the famous bacillus, others, possibly, by organisms yet to be discovered. In one instance, certainly, an analogy to this supposition has become a demonstrated fact. For thirty years it has been known that the disease called anthrax or charbon is characterized by the presence of a large bacillus ; yet in some cases the site of inoculation was indicated not by a malignant pustule or carbuncle, but by a local necrosis with subcutaneous formation of gas. These cases were designated charbon symptomatique. Five years ago Bollinger discovered that the bacillus found in the so-called charbon symptomatique is another variety than the bacillus anthracis which characterizes the malignant pustule—distinguished by both morphological and physiological features. And now the two diseases are recognized as etiologically distinct, though anatomically and clinically almost identical. To-day we can say with Schottelius, that *one* infectious disease, one infectious *tuberculosis*, is characterized by the presence of Koch's bacillus, though there may be others, clinically and anatomically entitled to the same name, which future research may distinguish etiologically from *this* tuberculosis, just as charbon symptomatique has been distinguished from charbon. Indeed, some observers have already expressed the suspicion, based on their own investigations, that there is more than one bacillus tuberculosis.

The association of Koch's bacillus with tuberculous tissues, and its absence from other structures, is therefore demonstrated and acknowledged; and this fact, taken in connection with Koch's own demonstrations, constitutes an array of evidence which has induced numerous German, Austrian, and English pathologists to accept as a fact the vital activity of the bacillus as the starting-point of the disease. Among these is Billroth, whose acquiescence is notable not merely because of his eminence as pathologist and surgeon, but because his own elaborate researches upon bacteria, published in 1874 and still widely quoted, led him to the conclusion that these organisms appeared in human tissues as the result, and not as the cause of morbid processes. Billroth, like the German pathologists generally, is open to conviction.

Yet while we may have, probably have, found in this

bacillus the object whose presence is followed by tuberculosis, we may not forget that the appearance of the disease implies not only the presence of this organism, but also the existence of animal tissues which permit the bacillus to exercise its vital functions. Many animals, even some rabbits, resist inoculation with the freshest tuberculous material. There is, in other words, a predisposition of the animal—an adaptation of his tissues favorable to the growth of this organism. The palm-tree cannot grow in Greenland; the oak does not flourish in the desert; the bacillus anthracis and the bacillus tuberculosis rarely grow in the body of a dog. And it may not be forgotten in the excitement over Koch's discovery, that there remains much to be done in determining the nature of this predisposition *of the animal soil to the growth of the tuberculous plant.* Thus far we are utterly in the dark. Dr. Formad thinks he discovers a ray of light issuing from certain narrow lymph-spaces. If he will prove what he asserts, he will have made a valuable anatomical contribution; yet when we remember that rabbits and guinea-pigs are peculiarly susceptible not only to tuberculosis but also to anthrax, and that cats and dogs are as markedly insusceptible to the one disease as to the other, it becomes evident that there must be some factor in the common predisposition to both diseases alike, which is not visible in the field of the microscope.

Indeed, with all due honor to Koch, and admiration for the most brilliant of experimental researches, we must admit that the discovery of the bacillus has chiefly an anatomical value: it localizes in this organism the infectious principle which had long been known to exist; it enables us to distinguish—ante- and post-mortem—infectious tuberculosis from inflammation, tubercular or other, due to other causes; but it does not as yet explain the hereditary predisposition, nor why this infection occurs in one man and not in another exposed to the same influences.

A dozen questions should be considered in this connection—the etiological identity of scrofula, tuberculosis, and fungus joint granulations; the possible infection of the infant by the mother's breath and breast, by the application of a handkerchief to the child's nostrils, etc.

Yet time permits a reference, and that but brief, to one of the most important—the possibility of infection from tuberculous meat and milk. For the so-called pearl disease of cattle, while presenting certain histological differences—an excess of calcareous salts, ascribable to their vegetable food—from tuberculosis in man and other animals, would seem to have an identical etiology; since inoculation with minute pieces into the anterior chamber of the rabbit's eye gives precisely the same result—local and general tuberculosis—as is induced by the same quantity of human tuberculous tissue and by nothing else. The effects of introducing the two into the circulation are also identical. Indeed there now remain but few who are not satisfied of the etiological identity of the two processes, especially since Koch's bacillus is found to inhabit both tissues. Yet etiological identity does not prove the possibility of infecting the human subject with tuberculous meat and milk. For it is a principle that must be borne in mind—a principle which Pasteur, in his ideas of preventive vaccination seems to have forgotten, by the way—that a material which can infect a given animal when placed in the eye, may fail when introduced into the alimentary canal or even under the skin. Koch found, five years ago, that although a mouse is so susceptible to anthrax as to be a reliable reagent in testing the strength of anthrax material when introduced subcutaneously, yet all attempts to induce the disease in mice, as well as in rabbits, by feeding them with anthrax tissues or spores, were quite unsuccessful. Since anthrax bacilli grow best in a somewhat alkaline liquid, and not at all in one markedly acid, the explanation may lie in the general acidity of the gastric and intestinal secretions in certain animals. But whatever the explanation the fact remains.

Many experiments have been made to determine the possibility of infecting animals by feeding them with tissues and milk from tuberculous cattle. Gerlach, Orth,

Bollinger, Klebs, and Chauveau were almost invariably successful with herbivorous animals; but Colin, Günther, and Müller saw only negative results. Virchow, experimenting with pigs, achieved somewhat indecisive effects; and while not inclined to deny the identity of the two diseases, he thinks it not yet experimentally proven that tuberculosis can be induced in animals by feeding them with such meat and milk. While we may quite agree in this, yet when we consider the *probability* on anatomical and experimental evidence; when we remember the peculiar frequency of intestinal tuberculosis in infants, especially in those artificially nourished; when we think that thirty per cent. of certain herds of cattle are, according to Professor Law, demonstrably tuberculous; we may be inclined to dispense with further direct experimental evidence, and avoid such meat and milk.

## LECTURE IV.

In this discussion I have referred to various bacteria as distinguished into species by essential differences of form and function. In these latter days it has become fashionable to speak of these minute organisms as transient modifications, due to incidents of their environment, of one and the same organism. Nägeli, indeed, would include not only bacteria, but also some of the higher fungi in this hypothesis. As this seems as yet a speculation, based not so much upon direct demonstration, as upon deductions, it will not require discussion here.

An essential element of this theory, however, the so-called accommodative cultivation of bacteria, seems to be supported by certain experimental evidence. This assumes that the physiological characteristics may be modified by contact with unusual influences—by a change of environment, in other words—as to render the descendants of a given bacterium which is capable of successful contest with the living animal tissues impotent to maintain such combat; and conversely to confer upon a previously harmless bacterial species the power to invade and destroy a living animal. This hypothesis is so fascinating, the solution of many difficult problems is rendered thereby so simple, the reconciliation of conflicting observations and opinions becomes so easy, that every man becomes at once his own bacteriologist. Diphtheria, on this hypothesis, is not due to a specific bacterium, but to some of those usually guileless organisms which ordinarily inhabit the healthy throat, incapable of harm; but which, excited into unusual and perverse activity by unknown influences of atmosphere, etc., invade the body with disastrous results. The application of this assumed principle is evidently limited only by the fancy and ingenuity of the individual; we have been already amply entertained by theories ascribing typhoid fever to the hypothetical tonic influence of sewer-gas upon the bacteria inhabiting the alimentary canal, etc.

The evidence in support of this hypothesis consists of deductions by analogy and of experimental observations. Since the life of an organism is the resultant of many forces, it is *à priori* evident that a modification of one or more of these forces may be followed by a change of the resultant life. In the higher plants and animals we have abundant evidence to this effect; the domestic pigeon and the dahlia are examples rendered familiar to us by Darwin. Such modifications, it is true, require time; but in biology time is measured by generations, not by years; and since from one bacterium a second may be produced in thirty to sixty minutes, it is evident that a day may induce in these organisms the effects of a thousand years in man. The evidence by analogy with higher organisms supports then the theory in question.

This same principle—modification of function by changes of environment—which Grawitz and Buchner had vainly attempted to demonstrate, seems to have been demonstrated by Pasteur in his studies upon protective vaccination. He asserts that the microbes which he regards as the morbid agents in chicken-cholera can be deprived of their virulence by successive cultures in contact with air; so that a given quantity of such culture fluid causes effects far less severe than the same quantity before such modification. Pasteur subsequently applied the same principle to the mitigation of anthrax virus; indeed the list has been still further extended by himself and others. Since in all these cases the same general principle is illustrated, it will suffice for our present purpose to consider the mitigation of anthrax virus for the preventive vaccination of sheep.

Pasteur's theory is this: the anthrax rods, as found in the blood of an animal dead of the disease, when placed in a suitable liquid maintained at a temperature of 42–43° C. grow as usual into threads, but do not produce spores. After a certain time their vitality is lost; when transferred to another flask, kept under the same conditions, they do not grow nor reproduce. But at any time previous to this final extinction of vitality, the bacilli still exhibit life, though their ability to invade a living animal, *i.e.*, their malignancy, is diminished. There occurs, indeed, a gradual *diminuendo* of malignancy, their morbid effect upon an animal decreasing with the prolongation of their exposure to these conditions, high temperature and exclusion of oxygen, until finally both life and malignancy are extinguished. Pasteur found that after eight days the bacilli had lost their fatal power to destroy rabbits, guinea-pigs, and sheep, though these animals are peculiarly susceptible to this virus. He claims that he has thus mitigated the virulence of these bacteria, has induced a modification of function.

As to the accuracy of Pasteur's observation in this case there can be no doubt; the vaccination of thousands of animals has already proven that the mortality induced by such anthrax cultures is much less than that following the usual inoculations with fresh virus. But his explanation, that the decrease of malignancy is due to modification of physiological function, is a by no means necessary conclusion, since precisely the same results can be and have been secured, under circumstances which preclude the possibility of a transmissible physiological modification.

First among these methods is simple dilution. It has been long since and often demonstrated that the effect induced by the incorporation of these virulent organisms into an animal depends, *cæteris paribus*, upon the number introduced. Chauveau found that sheep which had survived injections of fifty to six hundred anthrax bacilli died after subsequent injections of one thousand bacilli each. Oemler had previously made analogous observations upon horses; Löffler upon rats; it is indeed an accepted principle that the effect of inoculation increases with the number of injected bacilli. The somewhat general impression that quantity exerts no influence upon the result, except as to time, may be true when the effect is manifested upon an inert, unorganized mass, but not in the case of a living animal.

Diminution of malignancy can be secured in other ways also, which seem to accomplish practically the same result, dilution of the virus. Nocard and Mollereau found that anthrax virus is attenuated by simply mixing it with twice its volume of oxygenated water under pressure. Four hours' contact produces Pasteur's premier vaccin (for the guinea-pig), ninety minutes' exposure the second. Chauveau makes the premier vaccin by exposing anthrax blood to a temperature of 50° C. for fifteen minutes; and the second vaccin by the same exposure for nine to ten minutes. Since oxygen under pressure, as well as a high temperature, destroys the anthrax bacilli, it would seem that these methods accomplished merely a dilution of the virus by killing a certain number of the contained organisms; for in the brief time required in these experiments a physiological modification seems scarcely possible.

According to a communication presented by Bouley to the French Academy of Medicine, Peuch discovercd that the effects of tag-sore virus (variola in sheep) decreased by simple dilution with distilled water.

In the case of chicken cholera also, the characteristic organisms of which have been "modified" by Pasteur through a long and interesting process, there is reason to suppose that this modification may be simply a dilution. For vaccination against the disease has been successfully practised by simply introducing into the animal a piece of blotting-paper on which the blood of an infected animal has dried. The bacteria in dried anthrax blood die in a few weeks, but those still living at a given moment exhibit their original functions if transferred to a proper soil, and it seems probable that the diminished virulence of dried chicken cholera blood is due to the death of some of the contained organisms rather than to a physiological modification. This belief is strengthened by a letter recently received from Dr. D. E. Salmon, of Ashville, N. C., in which he says: "I have vaccinated fowls experimentally, both by Pasteur's method and by a method of my own. My method is simply inoculation with a very diluted virus; when sufficiently diluted it produces only a very circumscribed, local irritation that does not affect the general health in the least. One or a dozen germs of this fatal disease may be introduced in the tissues and are unable to produce any effect whatever. Twenty, fifty, or a hundred, according to the susceptibility of the fowl, will produce a slight local irritation. Pasteur's method requires five to nine months to attenuate the virus; by mine it is accomplished in as many minutes."

The effect can be secured, therefore, by simple dilution of anthrax or chicken cholera virus, as well as by Pasteur's cultures, and there are other reasons for suspecting that his mysterious method for the mitigation of bacterial virulence is practically a dilution of the culture, or rather, of the contained bacteria. If Pasteur would demonstrate that his tamed bacilli transmit their tameness to subsequent generations, the question would be finally settled; he asserts, indeed, that he has observed such transmission "in a few cultures," but gives no particulars, while the extensive vaccinations already performed on sheep prove that even his first (weak) vaccin sometimes kills an animal. It is but just to state that Koch has recently expressed his conviction that a genuine physiological modification does occur in Pasteur's cultures; whether this conviction is based upon personal observation or not does not appear.[1]

Although proper functional activity may doubtless decrease susceptibility to infectious diseases, whether of bacterial or of still unelucidated origin, it is evident in our daily observation that such activity does not necessarily confer immunity. At present but two avenues to such acquisition are known, a natural attack, and the artificial induction of the disease in mitigated form. The immunity secured by one attack of variola, scarlatina, measles, whooping-cough, etc., and by artificial inoculation with variola, as was formerly extensively practised,

[1] I have not succeeded in procuring Koch's monograph; the above statement is taken from reviews of it in French and German journals, the *Deutsche Med. Wochenschrift*, and the *Revue Scientifique*, especially.

prompted experimentation in regard to other diseases. Inoculation of cattle with material from animals dead of infectious pleuro-pneumonia—lung plague—was begun in Holland in 1852, and soon extended to Germany and Russia. In Saxony the mortality, previously twenty-five to thirty per cent. of the herds, became ten, six, two, even one per cent. At the Cape of Good Hope, where seventy to eighty per cent. of the cattle died of this infection, the disease almost vanished after inoculation was extensively practised. The tag-sore of sheep was always robbed of many victims by artificial inoculation. But of the diseases of whose parasitic origin we have conclusive or strong presumptive evidence, every one may occur more than once in the same subject. It is evident therefore, first, that immunity against an infectious disease, in the ordinary sense of the term, implies not necessarily the absence, but merely a relatively slight degree of susceptibility; second, that the question must be studied as to each disease independently of all others.

Although the question of protective vaccination has been experimentally studied as to anthrax, charbon symptomatique, chicken cholera, septicæmia, by Chauveau, Toussaint, Semmer, Colin, and Rosenberger, yet the results are so closely associated with Pasteur's name and with anthrax that I shall omit extended reference to the pioneer workers and works, and consider as the most favorable example, the well-known experimentation in protecting sheep against anthrax by inoculation with the cultivated bacilli. This method of Pasteur, I might say, is the first one which has afforded results at all satisfactory; and the principle differs from that employed in lung-plague, tag-sore, etc., in that the artificially cultivated organisms isolated from the accompanying animal tissues are employed—a new departure therefore.

In considering this subject, with which Koch's name is almost as closely associated as Pasteur's, it is advisable again to remember that this is a question of facts and not of individuals; that to us Gaul and Teuton are alike friends, as we fortunately keep no watch on the Rhine;

that neither Pasteur's brilliant work on fermentation, nor Koch's services on anthrax and tuberculosis; neither the unreasoning enthusiasm of the French for Pasteur, nor the intelligent confidence of the Germans in Koch; neither the grandiose egotism and artful dodging of the former, nor the apparent personal rancor of the latter; none of these may obscure our vision in estimating the value of present evidence.

Pasteur's theory may be briefly stated as follows: Since anthrax does not recur in the same individual, immunity against it as against other infectious diseases may be secured by one attack; the same effect may be obtained as in the variola of the human subject, by a harmless inoculation with the specific virus after exposure to unusual influences whereby its effect upon the animal is diminished.

To this theory Koch remarks that although some of the infectious diseases occur in the same animal but once, as a rule, yet no immunity is secured from others by the first attack; and adduces erysipelas, the septic diseases, gonorrhœa, intermittent fever, and recurrent fever, as examples familiar to all; the last-named is especially interesting, because it is invariably associated with a specific bacterial form—the spirochaete of Obermeier—though final proof of the causal relation of the parasite has not yet been furnished. But more than that: Koch points out, by the records of Prussian veterinary surgeons, that anthrax itself not infrequently occurs twice in the same individual; instances Oemler, who experimented on about one hundred animals years before Pasteur began to work upon the subject; and who saw horses, for instance, exhibit all the symptoms of anthrax once, twice, even eight times—at intervals of weeks or months, after inoculation with anthrax material'; quotes Jarnowsky, who saw the disease occur among fifty human patients, twice in one at an interval of two years; three times in another at intervals of two and three years. Löffler found that of 52 rats which were inoculated at intervals of some days or weeks, with the fresh virus, 30 survived the first, 23 the second, 13 the third, 3 the fourth, and 1 the fifth and sixth inoculation. Koch reminds Pasteur, therefore, that even though an animal survive a virulent inoculation he is not thereby secure against subsequent infection with anthrax. Further, Koch calls attention to the fact—proven by himself and others—that immunity against subcutaneous inoculation is not necessarily synonymous with immunity against infection through mucous membranes, especially of the alimentary canal; although Koch admits that as the fact had been proven only for horses, dogs, mice, rats, and rabbits, it might be otherwise with regard to sheep, with which he had not at that time experimented.

Such were the considerations advanced on either side.

Pasteur's theory was soon extensively tested—in Prussia and Hungary the experiment was superintended by an official commission of medical officers of the government. The proceeding has been usually the same. Pasteur vaccinates first with a weak virus, two weeks later with a stronger one, and after two further weeks, the animal is considered protected; and those thus protected, as well as others not vaccinated, are inoculated with material fresh from an animal dead of anthrax. At the end of the experiment in Hungary, fourteen per cent. of the protected animals were dead—mostly in consequence of the second protective vaccination; ninety-four per cent. of the non-vaccinated died. In Prussia the result was more favorable: 3 out of 25 sheep (twelve per cent.) died after the second protective vaccination. After the final inoculation with fresh blood, all of the non-vaccinated, but not one of the vaccinated, died. Pasteur thus demonstrated that sheep at least may acquire increased power of resistance to subcutaneous inoculation with anthrax; but he demonstrated at the same time that his protective vaccination destroyed almost as large a per cent. of animals as usually die from spontaneous infection in the pasture. Since that time Pasteur seems to have employed less virulent material, for according to accounts in French journals the mortality from the protective vaccination has been often only three, two, one, or even less per cent. But as Koch very properly observes, the ability of an animal to withstand a mild inoculation is not the question at issue, for that has been long known; the subsequent power to resist virulent material is the mooted point. The effect of Pasteur's own virus seems by no means uniform, since Duclaux, his assistant, who probably had virus of the proper attenuation, lost 20 out of 80 sheep in one flock during the two weeks after protective vaccination, and 11 out of 60 in another; yet in a third flock of the same race he lost only 1 out of 50. In the session of the Paris Veterinary Society, June 8, 1882, it was announced by Weber that 23 out of 993 sheep (2.3 per cent.) had succumbed to the preventive inoculation; at a later session Mathieu reported 29 deaths among 896 vaccinations (3.2 per cent.). At Salzdahlum 2 of 82 sheep died of anthrax after the second vaccination; in Kapuwar 5 of 50, and in Packisch (as already stated) 3 out of 25. Oemler lost 26 among 703 (3.7 per cent.). Dr. Klein recently called the attention of the British Government to the fact that Pasteur's vaccine virus was on sale in England; that he (Klein) had found that even the first and weaker virus could kill animals, having himself lost two sheep by such vaccination. Pasteur replies that Klein must have allowed other bacteria to invade the anthrax liquid.

From these accounts it would appear that Pasteur's preventive inoculation is a somewhat perilous performance, since even when performed by his own assistants it has killed 10, 12, even 25 per cent. of the vaccinated animals. Yet Pasteur recently stated that of nearly eighty thousand sheep vaccinated in France during the past year or two, none had died of the preventive inoculation.

And now for the second question: Does Pasteur's vaccination protect the sheep which survive it against anthrax? Against subcutaneous inoculation it certainly does for some weeks, as demonstrated in Prussia and Hungary; how long this protection endures and—more important still economically—whether it protects against infection through the mouth and alimentary canal, the usual mode of infection in the spontaneous anthrax of the pastures—these questions are not yet decisively answered. Yet we have already some data, collected largely by Pasteur on the one side, and by Koch on the other. Boutet reports that in the department Eure-et-Loire (where anthrax is especially prevalent, according to Pasteur) the general mortality last year was three per cent.; while of 79,392 vaccinated animals only .7 per cent. died; one herd of 2,308 vaccinated sheep lost only 8, less than .4 per cent.; another of 1,659 unvaccinated animals exposed to the same conditions lost 60 (3.6 per cent.). Nocard vaccinated in August, 1881, 380 sheep, reserving 140 for comparison; during the following five months 4 of the former and 15 of the latter died of spontaneous anthrax. According to these reports, therefore, the mortality of the vaccinated is only about one-tenth that of the unvaccinated animals, though it must be remembered that this does not include the mortality from the vaccination itself.

Other results are decidedly less favorable. Of 266 sheep vaccinated in Packisch last spring by Pasteur's assistant, Thuillier, there died between May and November, of spontaneous anthrax, 4; of 243 unvaccinated kept for comparison, 8. Cagny reported to the Paris Veterinary Society the following observation: In 1881, 20 sheep were vaccinated *à la* Pasteur; in February, 1882, 10 more. They belonged to a herd of 250, 2 of which died of spontaneous anthrax in May and June, 1882; it was therefore deemed advisable to revaccinate, which was done in July with Pasteur's stronger virus. Of the 20 animals vaccinated in 1881, none died of this inoculation; but of the 10 vaccinated only five months previously, *nine died;* of 4 non-vaccinated animals inoculated for comparison at the same time, 3 died. Koch vaccinated 8 sheep after Pasteur's method, and then inoculated them with virulent material; one died. The remaining 7 had therefore been inoculated with anthrax three times.

Twelve days later material containing anthrax spores was mixed with their food; 2 of the 7 died of anthrax.

A review of the evidence already adduced indicates that although Pasteur's theory has been demolished beyond repair, yet he has established the fact that the susceptibility of sheep to anthrax can be diminished by vaccination with the cultivated bacilli. From the economic standpoint the value of this measure is still debatable, though the aspect of the question has certainly improved since 1881, when the Hungarian medical commission, after inspecting Pasteur's experiment, advised their government not only to withhold its official sanction, but also to forbid all private experimentation. Nocard and Mollereau claim to have secured immunity of guinea-pigs by their method with oxygenated water; Chauveau with heated anthrax blood. If these claims be substantiated we may hope that simple dilution may furnish a virus which shall protect against other diseases as well as against anthrax. At any rate, Pasteur has established a principle, and it is to be hoped will be merely stimulated by the present unsatisfactory results to future and more successful efforts for its application.

Having thus sketched the present state of knowledge as to the agency of micro-organisms in the induction of disease, I may be permitted to consider briefly certain deductions therefrom, collectively known as the germ theory. I would emphasize the remark that the facts already summarized are not to be confounded with the speculations in which it may please one or another of us to indulge. Ocular demonstration removes a chain of events from the realm of speculation to the domain of fact; hence the germ theory has no longer jurisdiction over anthrax at least. A possible future demonstration that small-pox, syphilis, etc., are not caused by micro-organisms would be disastrous to the germ theory, but could not change the facts already established as to the morbid agency of the bacillus anthracis.

The germ theory supposes, if I understand it aright, that all infectious diseases are caused by the vital activity of parasitic organisms. In support of this theory there is certainly strong presumptive evidence: the stage of incubation; the unlimited reproductive power of the virus; the cyclical course and self-limitation of the disease. The stage of incubation can be explained by the assumption of no unorganized virus; all mere chemical compounds with which we are acquainted, even the ferments ptyalin and pepsin, begin to manifest the characteristic effects as soon as absorption has occurred.

Panum found that even boiled putrid materials, *i.e.*, the products of bacterial activity, though inducing the other features of septicæmia, failed to exhibit this characteristic incubation. By assuming an organism as the infecting agent this phenomenon becomes intelligible: the stage of incubation is then the period during which the inducted organisms are multiplying. In this way, too, the various durations of the incubative stage characteristic of the different diseases become intelligible. We cannot conceive that one chemical poison should require two to four days for the manifestation of its constitutional effects, as in scarlet fever, another forty days, as in syphilis; but we know that different micro-organisms multiply with different degrees of rapidity. A micrococcus may produce a second in thirty minutes; the bacillus anthracis may accomplish its entire vital cycle in twenty-four hours; the bacillus tuberculosis seems to require days. The unlimited reproductive power of the virus, characteristic of many infectious diseases, cannot be attributed to an unorganized poison or even organic ferment. A drop of blood from an animal poisoned with opium or strychnine exhibits only the power of the diluted poison. A drop of vaccine lymph, of variolous or gonorrhœal pus induces, in successive generations, unlimited quantities of identical materials. This effect cannot be justly ascribed even to any physiological unorganized ferment—a favorite refuge of those who are determined to deny to bacteria any influence whatever. Ptyalin can, it is true, convert into grape-sugar many times its bulk of starch, and the ptyalin is not thereby diminished in quantity; but it is not increased. No more perfect illustration of this principle can be furnished than an experiment of Rosenberger, which has especial value because this observer would assign to bacteria a subordinate rôle in morbid processes. He found that the boiled blood and tissues of septic animals—proven to contain no living bacteria—induced septicæmia as certainly as the same blood unboiled. He then placed in one of two flasks containing identical culture-liquids a few drops of the boiled blood, and in the other the same quantity of unboiled blood. Two days later every drop out of the latter flask conveyed septic infection, while large quantities from the former induced no reaction; every drop of the septic fluid was swarming with bacteria; in the other flask there were no organisms.

The virus of an infectious disease must then be something capable of reproduction, and this power is the peculiar characteristic of an organism. No unorganized poison, acid, salt, alkaloid, ferment is at present known which is capable of manifesting the phenomena shown by the virus of syphilis, variola, scarlatina, etc.

Turning to diseases whose parasitic origin is already demonstrated, we find all the characteristics of the infectious diseases exquisitely exemplified. Anthrax is marked by a stage of incubation—twelve to seventy hours—during which the bacilli multiply and effect access to the blood; the onset of constitutional disturbance is marked by the presence of numerous bacteria; with the death and disappearance of these, convalescence begins; the disease is eminently communicable by contact and yet may occur also sporadically and epidemically.

And this leads me to mention a fact often urged as an objection to the parasite theory: that many infectious diseases are intimately associated with climate, soil, and topographical features, indeed indigenous to certain districts. This is, in fact, strongly favorable to the germ theory. Many plants and animals of larger growth have decidedly limited habitats; botanists and zoölogists have long since informed us that this same principle applies to microscopic organisms, including fungi; shall bacteria then, a family of fungi, be exceptions to the general rule? As to anthrax, the case is answered; the disease is endemic in certain districts, that is, the bacilli grow outside of the animal body only in these districts. Koch has recently endeavored to elucidate the reasons for this. He ascertained that the growth of the anthrax bacilli requires moisture and a temperature of 15° C. It is evident, therefore, why anthrax is not endemic in districts whose surface temperature fails to reach this point. Koch further ascertained, from the official reports of Prussian veterinary surgeons, that after the overflow of rivers and lakes an outbreak of anthrax had been frequently observed in cattle pasturing at certain points along the banks. He found by experiment that the bacilli flourish in infusions of various grasses, grains, and vegetables. Hay infusion is usually a poor soil, because of acid reaction; when rendered neutral or slightly alkaline, the bacilli grow in it luxuriantly; but it had been long before stated by German and French observers that the anthrax districts usually had a calcareous soil. Hence he conjectures that in such districts the alkalinity due to the lime may render even hay a soil favorable to the anthrax parasite; that it may usually grow on decaying plants in such districts. That anthrax is especially prevalent in autumn seems to result in part at least from the fact that these bacilli, like many other fungi, grow only on dead plants. It might be interesting to review for comparison with anthrax the facts which establish the association of various infectious diseases—cholera, yellow fever, the malarial fevers, for example—with local influences of soil and temperature; the origin of typhoid fever in particular wells and springs, as has been conclusively established by observations upon the German and Austrian soldiery. Yet such discussion would transcend my time and my province, since I have attempted to portray what has been, not what remains to be accomplished.

Certain popular arguments against the morbid agency

of bacteria are worthy of consideration perhaps, though not because of their intrinsic weight. To some it is incredible that bacteria should harm us, since we live in health though surrounded by them—eating, drinking, and breathing them. If it be remembered, however, that the name bacteria is merely a convenience for designating organisms of widely different functions, this argument seems less formidable. On the same principle it might be asserted that all mammalia are harmless because we come into daily contact with sheep, cattle, horses without injury. The diversity in function, food, etc., among the microscopic beings is not less marked than among the larger organisms: there are bacteria, and bacteria. It is surprising that Mr. Cheyne, in his admirable work on antiseptic surgery, falls into a similar error. He admits that bacteria are not infrequently found under Mr. Lister's own dressings; that sometimes the course of the wound appears thereby unaffected, while at other times he thinks he has observed that the wounds heal less kindly. He consoles himself, however, with the reflection that these are "only micrococci." He seems to forget that several species of micrococci are distinguishable in form, size, color, and function from one another and from all others. The fact that certain micrococci found during different diseases are morphologically indistinguishable from others found under Mr. Lister's dressings, does not justify the assumption that all are functionally identical. If Mr. Cheyne were requested to swallow some pills, he would probably inquire as to their contents before complying, and would not be satisfied with the assurance that they were "only pills." Yet we have every reason for asserting that the minute globules known collectively as micrococci present differences as great as the larger globules designated, for convenience, pills. When we reflect that the active agents in the induction of pyæmia are micrococci; that the organisms found in malignant diphtheria are micrococci, we must protest against Mr. Cheyne's promiscuous ascription of benign qualities to any tribe of bacteria, even if "only micrococci." There are micrococci and micrococci.

Again, it is said, how is it possible for recovery to occur from a disease caused by bacteria? What stops their growth? So far as I am aware, this question has not been decisively answered. Several facts suggest that the products of their own vital activity arrest further development. Analogous facts have been demonstrated: the mucor racemosus ceases to grow in a liquid when the alcohol produced by its own vital action exceeds a certain percentage, though there may still remain fermentable sugar in abundance. During putrefaction there are produced numerous compounds, of which one at least, carbolic acid, arrests, even in small quantities, further development of putrefactive bacteria. Brieger has recently shown that the infectious diseases proven clinically and experimentally to be caused by putrid infection — pyæmia, diphtheria, erysipelas — are distinguished by the excretion in the urine of excessive quantities of carbolic acid; while in other diseases exhibiting equally intense fever and constitutional disturbance—acute rheumatism and variola, for example—the amount of this acid in the urine is normal or subnormal. Hence the conjecture that the bacteria are both bane and antidote. Yet a failure to explain the phenomena satisfactorily does not, of course, impair the stability of the fact. The bacilli of anthrax are observed to become, in the living animal, pale, of uneven outline, incapable of absorbing staining fluids; in short, they are dead. With their death the convalescence of the host begins, as a rule.

It should be remembered that other parasites than bacteria may cause disease, some of them, perhaps, overlooked in the universal hunt after bacteria. Koch calls attention to Woronin's discovery that a disease of cabbages is caused by an amœboid parasite, which enters the root of the plant and becomes almost indistinguishable from the proper vegetable cells; and suggests the possibility that some of the amœboid bodies known as white blood-corpuscles in animals may be intruding organisms, especially since Ehrlich has shown that different leucocytes exhibit various reactions to staining agents. Five years ago it was discovered that a mould-fungus, the actinomyces, induces fatal disease of man and other animals; Wittich found organisms, which he calls spirilla, in the blood of apparently healthy gophers; Koch found numerous organisms (monads) in the blood of five gophers that had died without other discoverable cause. A new filaria has been recently discovered in the human subject by Bastian, and similar discoveries are reported in the camel and the hog.

There is probably no one among us who doubts that the trichina spiralis can and does induce in the human subject a serious, even fatal disease; yet the evidence as yet adduced is merely the association of the worm with the morbid condition, for no one, so far as I am aware, has ever induced the disease by introduction of the isolated worms. Yet the same men who assert the pathogenetic influence of the trichina, contemptuously reject the idea that leprosy, tuberculosis, recurrent fever, and pyæmia are caused by bacteria, although the evidence—constant association of the parasite with the morbid condition, applies to all cases. Indeed, the weight of evidence is decidedly in favor of the bacteria; for the trichina is found not only in the subjects of trichinosis, but also in many individuals who have never been suspected of harboring the worm. It is not extremely seldom that trichinæ are found in the bodies of patients who have died of acute disease, wounds, accidents, etc.; indeed, an examination of several thousand consecutive cadavers in German hospitals, some years ago, revealed trichinæ in over two per cent., without regard to the cause of death. It might, therefore be argued that the presence of the worm is a mere accident—an epiphenomenon, observed in healthy as well as in diseased conditions. The bacilli of leprosy, on the other hand, are found *only* in patients suffering from this disease. I would not express any doubt, by this comparison, of the morbid agency of the trichina, but would merely call attention to the fact that for this belief we have really no more conclusive evidence than we have for accepting the pathogenetic influence of bacteria in leprosy and in several other diseases. Yet it will doubtless be years before some of us realize the fact that in this unreasoning and prejudiced opposition to demonstrated facts, we are playing the unenviable rôle of the cow to the locomotive of advancing science. George Stephenson's prophecy—that the result of such collision would be "bad for the cou"—has been often fulfilled; neither horned cattle nor pseudo-bacilli have materially retarded the progress of science. The improvement of means and methods for minute investigation has ever been and must still be followed by further advance into the realm of the minute, whose boundaries doubtless stretch far beyond the present means of optical exploration. After centuries of controversy the intestinal worms banished the "vermicular diathesis;" the acarus scabiei conquered the "itch cachexia;" the step from the acarus to the bacillus tuberculosis is not quantitatively greater than from the tape-worm to the acarus; each is merely the measure of successive improvements in means for minute observation. Yet the same spirit which nursed the vermicular and itch diatheses, will doubtless for years see in bacteria only fibrin threads and fat crystals; and will cling with heroic devotion to the tubercular diathesis, to "micro-necrosis," and to narrow lymph-spaces.

When we glance over the progress of the last few years; when we consider the life-saving revolution in surgical methods; when we regard the enhancement, indeed the very salvation of enormous economic interests by the eradication of the bacterial disease of silkworms, without mentioning preventive vaccination against anthrax and chicken-cholera, we may search in vain the records of other departments of science during the same period for discoveries which have secured direct personal and pecuniary advantages comparable to those derived from our present incomplete knowledge of the relations of micro-organisms to disease. And when we consider the problems already half solved, the questions to whose solution the way appears open through the same methods already successfully applied to anthrax and tuberculosis, we may hope for results to which present knowledge shall seem a mere introduction. But these results can be secured only by earnest, skilful, continuous experimental investigation, which is practically impossible without pecuniary support. In France and Germany such support is liberally supplied by the government; in the United States, where human life is certainly as valuable as there; where live-stock interests are already greater than in these countries combined, and must multiply many fold in the immediate future; where a single infectious disease of cattle has caused the loss of $20,000,000 in one year, and a single disease of hogs the destruction of $30,000,000 in the same time; where infectious diseases are so prevalent among live stock that the fear of infection has closed European markets against American meat and cattle—the government of this great commonwealth, which advances enormous sums for local river and harbor improvements; which sends expensive commissions over the world to observe the transit of Venus or of the moon; to find an open Polar sea; and engages in other undertakings of purely scientific interest, has not yet made one judicious, systematic, liberally supported inquiry into the possibility of acquiring protection against pleuro-pneumonia, hog-cholera, and other devourers of the national wealth. A glance at the Imperial German Health Bureau and its work during the last four years, and a mental comparison of the pecuniary resources of Germany with those of the United States, inspire the hope that we shall not always lag so far behind in matters which appeal to the tenderest spot of the American anatomy—the pocket.

In concluding these lectures, Mr. President and gentlemen, I shall offer no apology for their fragmentary character, since I would not call attention to defects already amply apparent. Yet I venture to hope that one merit may be accorded them—that they constitute an impartial and unpartisan attempt to portray the present status of this vexed question. And this I hope, not on personal grounds, but because a suspicion of insincerity in the portrayal would retard acquiescence in what I must and do regard as truth.

If these lectures shall serve as a vehicle for conveying to the busy practitioner facts which he has not time to seek amid the mass of current literature; shall contribute, however little, to the more general discrimination between theories and facts, between observations and deductions, between assertions and demonstrations; shall tend to confirm the belief that much may be hoped for, though perhaps but little is already completed in this direction—their object will be accomplished, and your lecturer will hope that he was not inexcusably presumptuous in consenting to appear upon a platform which has been honored by a Bartholow and by a Dalton.

## APPENDIX A.

The following cuts are copied from some of the twenty-eight photomicrographs exhibited at the lectures. It has been deemed advisable to print them here with the briefest possible summary of the remarks which accompanied their exhibition.

In erysipelas the lymph-spaces just at and in front of the advancing edge of the inflamed area contain micrococci. These organisms have been isolated by cultivation on solid media (Koch's method) by Fehleisen. He reports the successful induction of erysipelas by inoculation with the micrococci thus isolated, in eight rabbits and in one human patient.

Recurrent or relapsing fever is characterized by the presence of Obermeier's spirillum in the blood.

As yet no successful inoculation with the isolated spirilla has been reported.

A short, thick bacillus is found post-mortem in the liver, kidney, spleen, and lymph-glands in the majority of cases of typhoid fever (Eberth, Koch, Friedländer); and larger bacilli in the vicinity of the intestinal ulcers (Klebs). Maragliano asserts the presence of both varieties in blood drawn by a hypodermic syringe from the spleen in fifteen cases (*intra vitam*). It would seem, however, that he protests too much; for even post-mortem only Eberth's, never Klebs' bacilli, are found in the spleen.

In croupous pneumonia small bacteria have been found at the edge of the advancing inflammation (like the micrococci in erysipelas); also in some internal organs. The same bacteria have been found in the lung immediately post-mortem, and even *intra vitam*.

Endocarditis ulcerosa seems to be one of the forms of pyæmia, sometimes of spontaneous (*i.e.*, undiscovered) origin.

The internal organs as well as the cutaneous tubercles of patients afflicted with leprosy contain a distinct bacterium—the bacillus lepræ. Attempts to induce the disease in the lower animals by inoculation with the isolated bacilli as well as with leprous tissue, have not as yet been successful.

Frisch, of Vienna, has found a characteristic short,

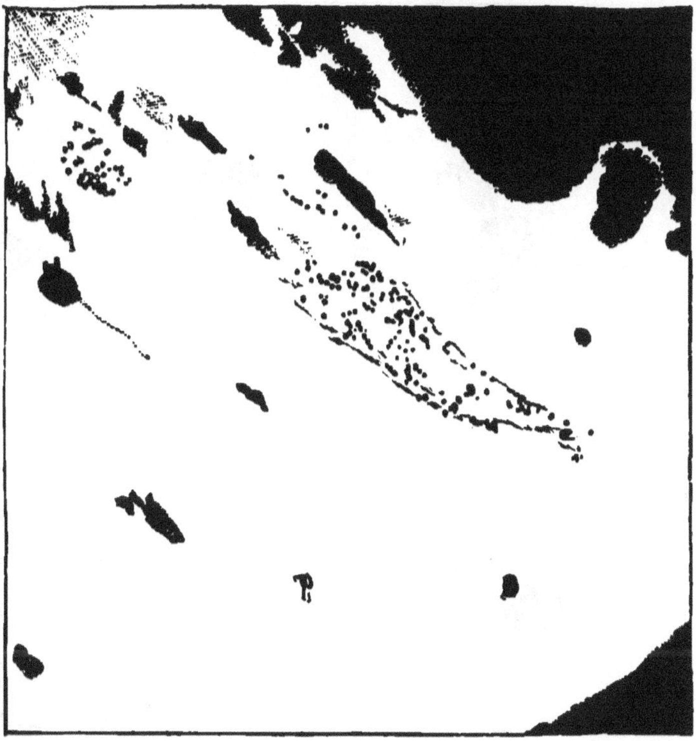

Fig. 9.—Skin excised, *intra vitam*, from a case ot erysipelas ; micrococci in lymph-vessel, × 700. (Koch.)

thick bacillus in every case of rhinoscleroma (twelve in number) which he has had opportunity to examine. Pieces were excised from the nose or mouth *intra vitam* ; the bacilli occur, like those of leprosy, in the large cells characteristic of the tissue. Frisch cultivated these organisms on solid media, after Koch ; but was unsuccessful in attempts to induce the disease in rabbits. It must be remembered that man is the only animal known to suffer from rhinoscleroma and from leprosy.

Klebs and Crudeli assert the induction of malarial fever in rabbits through the agency of a bacterium—the bacillus malariæ. Dr. Sternberg, U. S. A., has made a critical and experimental review of this work, from which

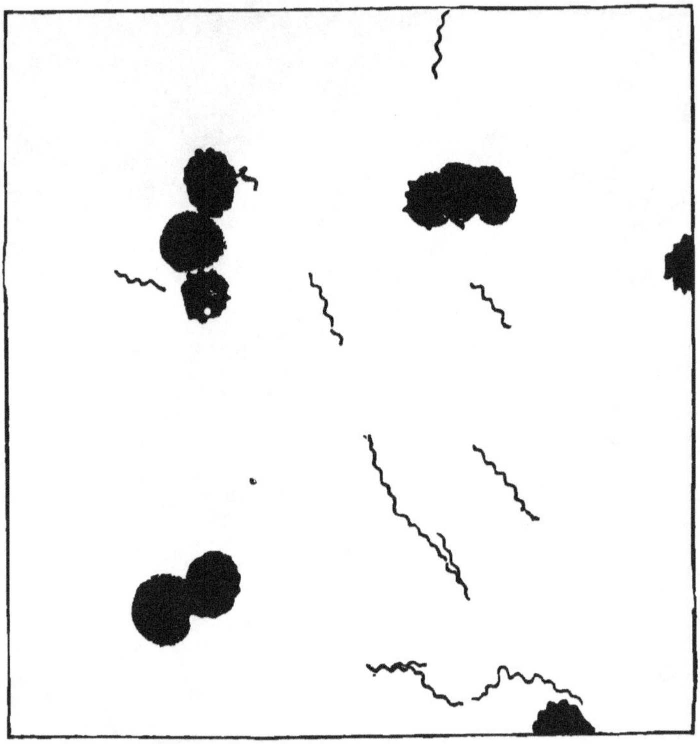

Fig. 10.—Spirochaete of Obermeier and human blood-corpuscles, × 700. (Koch.)

he concludes: "The evidence upon which Klebs and Crudeli have based the claim of the discovery of a bacillus malariæ cannot be accepted as sufficient," and "their conclusions are shown not to be well-founded." Such is the general opinion, so far as I have been able to ascertain, among those familiar with this department of investigation. Bacteria said to be identical with these have been found in the blood of patients suffering from malarial fever, and by Ziehl in one individual who had no symptoms nor history of intermittent fever, but was suffering from diabetes. In three of Ziehl's four cases, the bacilli disappeared from the blood after the administration of quinine for several days.

Bacteria of various kinds have been seen in syphilitic

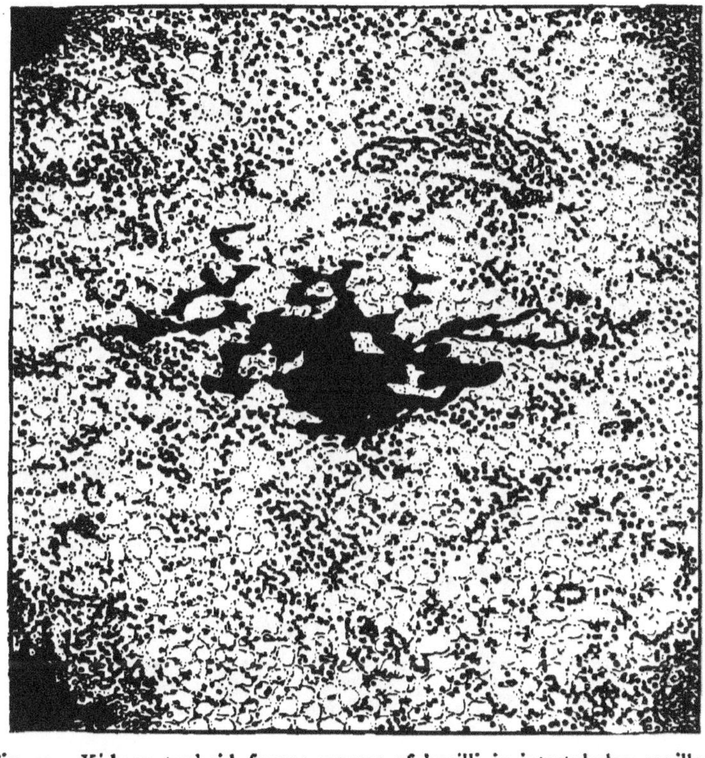

Fig. 11.—Kidney, typhoid fever; masses of bacilli in intertubular capillaries, × 100. (Koch.)

tissues, post-mortem, by different observers. Quite recently Birch-Hirschfeld announced the discovery of micrococci in twelve gummata, post-mortem; in three condylomata, one chancre, and one cutaneous papule excised during life. Morison, using aniline staining and Abbé illuminator, found bacilli in chancres and other syphilitic tissues—the same variety being present in all.

The contents of variolous pustules, like other pus often contains micrococci. No trustworthy observation of the presence of bacteria in the blood during this disease has been recorded. Ehrlich has sought them in vain even in hemorrhagic small-pox. Post-mortem they are sometimes found in the tissues.

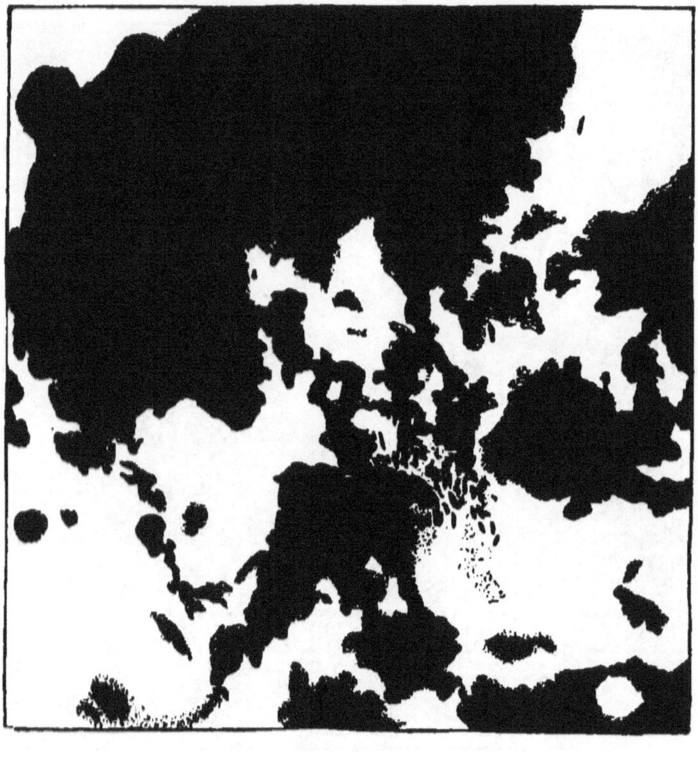

Fig. 12.—Edge of bacterial colony in the liver from typhoid fever; the individual bacilli are seen, × 700. (Koch.)

In diphtheria micrococci are often found not only in the local necrotic tissue, but also in internal organs and in the blood. In scarlatina no reliable affirmative observations have been made, so far as I am aware. But since diphtheria, scarlatina, and erysipelas must be classed according to clinical, experimental, and chemical (Brieger) evidence with the putrid diseases, the evidence in favor of the bacterial origin of septicæmia and pyæmia suggests analogous etiological influences for these diseases also.

Schütz and Löffler have recently reported the induction of glanders in rabbits and in two horses by inoculation with bacilli isolated by cultivation from animals suffering from this disease.

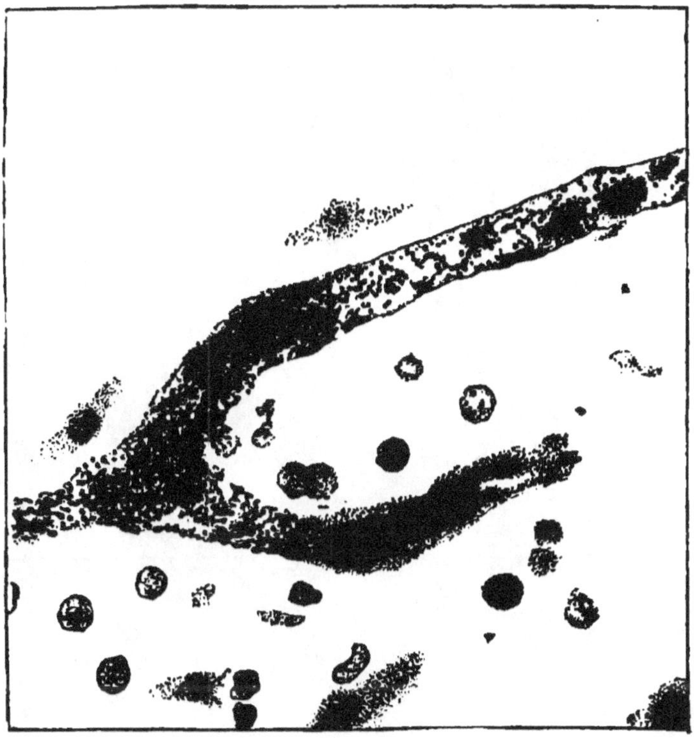

Fig. 13.—Intertubular capillary of kidney containing bacteria; from a case of croupous pneumonia, × 700. (Koch.)

Cuts of the trichina spiralis and of the filaria sanguinis are added as an illustration of the fact that morphological similarity does not prove physiological identity. These two nematode worms are, of course, easily distinguishable by the difference in size; yet structurally they are, in the larval state as here represented, quite similar. Yet the one is found coiled in the voluntary muscles, its migration from the intestine often causing symptoms simulating typhoid fever; the filaria, on the other hand, circulates with the blood (by night only, as a rule), and is associated with one of several morbid states—chyluria, lymph-scrotum, sometimes ending in pyæmia (as in the case which I was fortunate enough to observe in the London Hospital). In some

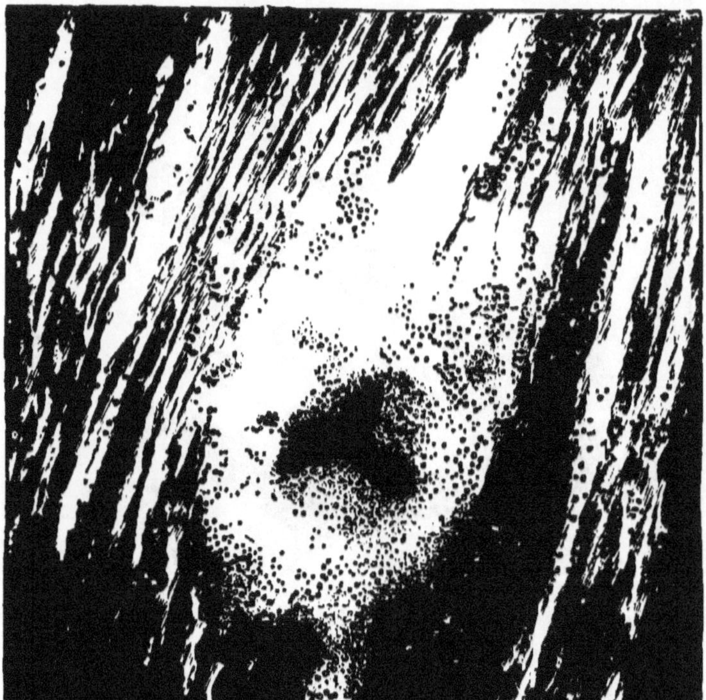

Fig. 14.—Incipient abscess formation around a vessel occluded by a micrococcus colony. Heart-muscle, endocarditis ulcerosa, × 100. (Koch.)

cases its presence seems to cause no abnormal symptoms.

One of the most interesting of recent observations is that of Brieger and Ehrlich (*Berliner Klin. Wochenschrift*, 44, 1882), in which they report two cases of Koch's "malignant œdema" in human subjects suffering

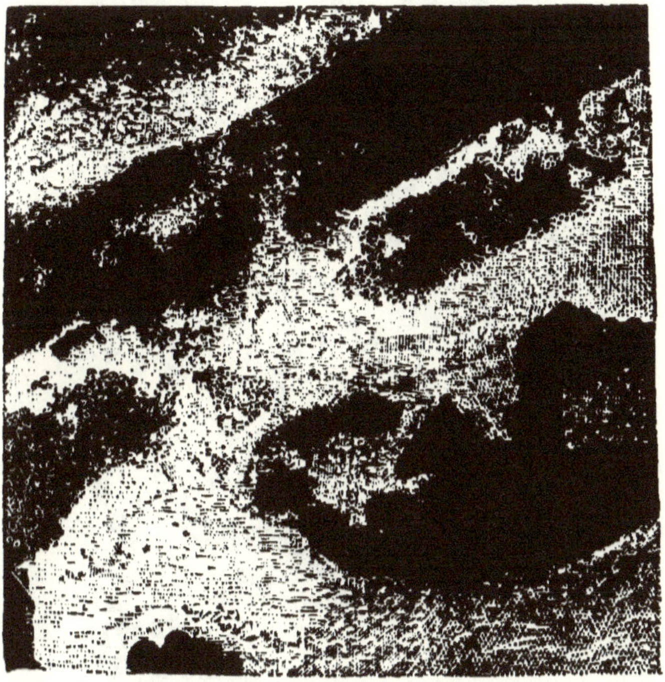

Fig. 15.—Micrococci in renal capillary, small-pox, × 700. (Koch.)

Fig. 16.—Trichina spiralis, × 50. (Oliver.)

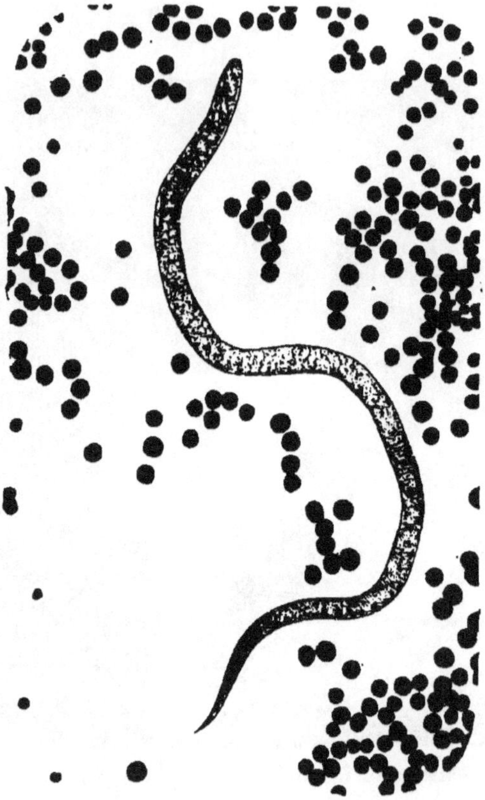

Fig. 17.—Filaria sanguinis hominis in human blood, ×280. (In this case from twenty-five to one hundred worms were found in every drop of blood between six P.M. and five A.M.)

Fig. 18.—Filaria, dried and stained to show the sheath (compare Fig. 16).

from typhoid fever. The bacillus characteristic of this affection is widely distributed, being often found in ordinary garden earth, and frequently appearing in animal bodies a few hours after death ; in these two cases the organisms seem to have been introduced by subcutaneous injections of musk. Koch and Pasteur have induced the disease in mice and rabbits by inoculation with earth and with the isolated bacilli ; and the morbid condition which often appears spontaneously in cattle, termed charbon symptomatique, or Rauschbrand, seems to be the same disease : but this is, I believe, the first instance in which the bacterium has been demonstrated in the human subject of the disease.

## APPENDIX B.

A sketch of methods especially adapted to the detection and recognition of bacteria is added.

The basic aniline pigments—Bismarck brown, gentiana violet, methyl blue and violet, and fuchsin—are especially valuable for this purpose, because readily absorbed and retained by bacteria (Weigert). These may be kept as simple filtered solutions (one to two per cent.) in distilled water, though the addition of a little alcohol (ten per cent.) is desirable to secure permanence and to prevent the growth of bacteria in the liquid—a possible source of error which must be always borne in mind.

Sections may be prepared from fresh or from hardened tissues, but must be *thin*, in order to permit satisfactory inspection of these minute objects; a microtome is therefore an extremely desirable, indeed almost essential, means for the preparation of such sections. For hardening tissues absolute alcohol should be used; small pieces —cubes not more than one-half or three-fourths of an inch in each dimension, for example—are sufficiently hard after immersion for one or two days in several (four or five) ounces each of alcohol.

After immersion for three to ten minutes in the two per cent. aqueous solution of the aniline color, a section exhibits an intense, diffused staining; it is then transferred for a few minutes to alcohol, which abstracts the color from all morphological elements (with certain occasional exceptions) except the nuclei of cells, and bacteria; it is usually desirable to transfer the sections to a second and even a third dish of alcohol, in order to wash off the coloring matter dissolved out of the section by the alcohol in the first dish. For permanent preservation the sections should be passed through oil of cloves and mounted in Canada balsam after the usual method;

for temporary inspection they may be laid in glycerine after washing with distilled water; but all except the brown colors gradually fade in glycerine. Bismarck brown is for this reason, as well as for the cleanness of its staining—*i.e.*, absence of granular precipitate—perhaps the most satisfactory for general use; though in special cases (tubercle bacilli and gonorrhœal micrococci for example) the blue and the red pigments are preferable.

The violet (methyl and gentiana) dyes are especially useful in the detection of amyloid degeneration; for while the normal elements are colored blue, those which have undergone the amyloid change exhibit an intense red color. This red tint rapidly disappears if the sections are placed in alcohol; hence they cannot be mounted in balsam, but may be kept (a certain time) in glycerine.

Bacteria which may be present in sections stained in this way can be usually recognized at once (especially when occurring in groups) by their intense color; yet their detection can be facilitated by taking advantage of the fact that these organisms not only absorb the aniline colors readily, but retain these colors in the presence of reagents which decolorize the other morphological elements that may be present: after exposure to such agents, therefore, the bacteria are distinguishable at once as the only intensely stained objects in the field. Occasionally other objects than bacteria are encountered which exhibit the same reaction—the plasma-cells of Ehrlich, nuclear detritus, globules of leucine. Ehrlich's cells, moreover, simulate groups of micrococci somewhat closely, but can be usually distinguished by the following characteristics: 1, the granules of these cells are rarely of uniform size, while the micrococci of a zooglœa mass are; 2, the granules composing the cell usually surround a clear, oval space—probably the location of the nucleus; no such appearance is presented by a group of micrococci; 3, numerous plasma-cells presenting a uniform appearance usually occur in the same section, even in the same field; they are seen not infrequently in tissues in a state of chronic inflammation.

Any one of several agents may be employed for this purpose of differentiating bacteria from surrounding objects by decolorizing the latter; simple immersion in alcohol or ether for some minutes sometimes accomplishes this result. The preferable agents are, however, 1, a solution of acetic acid (two to five per cent.), and, 2, a solution of the carbonate of potassium (one to three per cent.) in water. Immersion of the stained sections in one of these fluids for five to ten minutes usually suffices to decolorize the morphological elements other than bacteria, while the latter still retain a brilliant color. It must be remembered, however, that none of these rules is of universal application: different varieties of bacteria exhibit various reactions toward these agents, and some of the aniline dyes are less readily extracted from the nuclei than others; the blue and the violet are, in my experience, more tractable in this regard than the brown colors. When it is intended to differentiate in this manner the sections should be at first intensely stained, which may be accomplished by the use of stronger aniline solutions, or (preferably) by permitting them to lie for a longer time in the ordinary (two per cent.) solution.

The prominence of these colored bodies is further enhanced by illumination with the Abbé condenser, whereby refraction outlines are practically obliterated, the colored objects appearing alone—simply by virtue of their color—against a white background.

While the cutting of sections requires time and skill not usually possessed by the general practitioner, the preparation of liquids—blood, pus, sputum—for the detection of bacteria is extremely simple, and can be accomplished without special knowledge, skill, or apparatus. This is particularly fortunate, since a valuable means for clinical diagnosis is thereby brought within the reach of every one who possesses a microscope, even an inexpensive one. The points to be secured are: first, a very thin layer of the liquid under examination, and, second, the coagulation of the albuminous constituents of that liquid. A drop, a very small drop of pus, sputum, or blood is placed in the middle of a clean cover-glass; a second cover-glass is then laid upon this and the two are pressed gently together until the enclosed drop has become a very thin layer; the two cover-glasses are then separated by *sliding* one over the other, not by pulling them apart. The two glasses, each of which has now a thin layer upon its surface, are allowed to dry.; the coagulation of the albumen of the fluid (a most important measure, which prevents precipitation of granules from the staining fluid) is now accomplished by holding the cover-glass over a gas or alcohol flame until it acquires such a temperature as to be unpleasantly warm when applied to the skin; or the glass may be "passed three times through the Bunsen flame," as directed by Ehrlich. The pus or sputum is now stained either by floating the cover-glass on the staining liquid, or (more conveniently) by pouring a few drops of the aniline solution from a pipette onto the cover-glass; after five to fifteen minutes the staining fluid is poured off, the excess of color removed by washing in a stream of water; the cover-glass is now allowed to dry and is then ready for mounting in either glycerine or balsam. The methylene blue is perhaps the most satisfactory of the aniline colors for general use in staining these dry cover-glass preparations.

While most species of bacteria known to occur in human tissues are readily stained by these methods, certain varieties require special modifications thereof. Prominent among these is the bacillus found in tuberculous tissues, whose presence in sputum already possesses diagnostic (and prognostic ?) value. Of the various methods already published for staining the bacillus tuberculosis, Ehrlich's has the general preference, and has always given satisfactory results in the writer's hands. This method requires (*a*) a saturated solution of aniline oil in water,[1] made by adding the oil, drop by drop, to distilled water in a test-tube, which is meanwhile constantly shaken ; upon saturation, indicated by turbidity, the mixture is filtered. To the clear filtrate are added a few drops of (*b*) a saturated alcoholic solution of an aniline pigment, preferably fuchsin. Upon the appearance of turbidity (or opalescence) the staining fluid is ready for use (though it is by some observers thought desirable to filter before using). In this liquid sections of tissue or layers of sputum on cover-glasses (prepared as already directed) lie twenty-four hours at ordinary temperatures, or thirty to sixty minutes if the temperature be maintained at 50° C. At the end of this time all the elements, bacilli included, are stained ; in order to differentiate the bacteria the section (or cover-glass) is immersed for a few seconds in a mixture of nitric acid (one part) and water (three parts). The color disappears at once, and after washing the section in water and mounting in the usual way, it will be found that the *bacilli only* exhibit a deep color, the other objects in the field having little or no color. The staining is now complete—the red bacilli appearing against a white background ; yet a pleasing contrast may be obtained by coloring this background blue ; this is easily accomplished by immersing the decolorized section or cover-glass in the ordinary two per cent. watery solution of methylene blue. The already stained tubercle bacilli (red) are unaffected by the second dye and retain the first color ; while nuclei, etc., absorb the second (blue) color only.

The aniline oil should be fresh, for in time it becomes oxidized, a condition indicated by a dark instead of the original light brown hue ; in this condition the result of staining is uncertain. The bacilli are in some cases simulated in size and shape by minute crystals, but are, however, easily distinguished from the latter by the brilliant red color which the true bacilli exhibit (when stained

---

[1] Those who cannot conveniently procure the oil will find an excellent substitute in ordinary (pure) carbolic acid. The directions for use remain as above with the substitution of the words " carbolic acid " for " aniline oil."

with fuchsin). To secure thorough staining—of all bacilli present—it is advisable to expose the section to the staining-fluid for twenty-four hours.

Stained by this method the bacilli can be detected in sputum with an ordinarily good one-fifth inch objective (giving 250 to 300 diameters) and simple central illumination without condenser; though for critical examination, especially in tissues, higher powers (400 to 600 diameters) and special illumination are desirable. It is already demonstrated that the presence of these bacilli has decided diagnostic value, not only in sputum but also in pus from laryngeal ulcers, from cold abscesses, from joint cavities, and in urine.

The bacilli of leprosy may be stained in fresh sections with the ordinary two per cent. aqueous solutions of the aniline colors, without special preparation; if the tissue has been long exposed to alcohol it is desirable to place the sections for a few minutes in a five per cent. watery solution of acetic acid, and then transfer them to a ten per cent. solution of caustic potash. After washing in water they may be then stained with the ordinary fuchsin solution.

The micrococci of gonorrhœal pus require no special treatment; the cover-glass on which the pus is dried and heated (as already directed) is treated for five to ten minutes with the two per cent. solution of methylene blue.

For cutting extremely thin sections of lung and other delicate tissues, the following method of imbedding is well adapted:[1] The tissue lies for twenty-four hours in the ordinary clearing solution (one part of carbolic acid and four of turpentine); it is then transferred for twenty-four hours to a saturated solution of paraffin in turpentine, after which it may be imbedded either in pure paraffin or in one of the many mixtures of this substance with tallow oil.

---

[1] For a knowledge of this method, and of other valuable points in histological technique, I am indebted to the rare technical skill and extensive experience of my friend Dr. W. T. Councilman, Fellow (by courtesy) of the Johns Hopkins University, Baltimore.

www.ingramcontent.com/pod-product-compliance
Lightning Source LLC
Chambersburg PA
CBHW021938160426
43195CB00011B/1132
*9783337059590*